DATE DUE

MANAGING GEOGRAPHIC INFORMATION SYSTEMS

Managing Geographic Information Systems

❖

Nancy J. Obermeyer
Jeffrey K. Pinto

THE GUILFORD PRESS
New York   London

© 1994 The Guilford Press
A Division of Guilford Publications, Inc.
72 Spring Street, New York, N. Y. 10012

Marketed and distributed outside North America by Longman
Group UK Limited.

Printed in the United States of America

This book is printed on acid-free paper.

Last digit is print number: 9 8 7 6 5 4 3 2 1

Library of Congress Cataloging-in-Publication Data

Obermeyer, Nancy J., 1955–
　　Managing geographic information systems / Nancy J. Obermeyer and
　Jeffrey K. Pinto
　　　　p. cm.
　　Includes bibliographic references and index.
　　ISBN 0-89862-005-8
　　1. Geographic information systems.　　I. Title.
G70.2.O24　1994
910'.285—dc20
93-47199
CIP

To our spouses, Samory Rashid and Mary Beth Pinto, who have given us the love, support, and means to succeed in that which we attempt. This book is yet another example of what we have been able to achieve through their encouragement.

❖

Preface

When this book was conceived, the literature about geographic information systems (GISs) consisted largely of instructional texts (e.g., Stan Aronoff's introduction to GISs), readings in GISs (e.g., Peuquet & Marble, 1990), and scores of unrelated case studies with weak or—worse—nonexistent theoretical foundations. In fact, much of the case study literature is still located in proceedings and other fugitive literature. There was little available on managing GISs and even less that contained a strong theoretical component. This book is intended to fill that gap.

The approach to this book begins with the belief that theory and practice are necessarily linked. As Ostrom (1986) says, "Theory without experience is fantasy. Experience without theory is blind." Tobler (1976), a key figure in the development of both analytical and digital cartography (and GISs), reinforces the value of theory, making the point that whereas good theory typically has a half-life of some 20 years, technology has a half-life of only about five.

And yet, in the case of the literature on implementing and managing GISs, theory has been elusive. This situation has made it difficult to draw general lessons about implementation that may have widespread utility among those charged with the responsibility for putting in place and managing these systems.

It should not be surprising, however, that this situation exists. While the geographic information system as a technology has been around for a quarter of a century, the widespread implementation of the technology is much more recent. One would not, and indeed should not, expect literature on the theoretical aspects of implementing GISs to appear early in the development of this or any technology. Informed expectation would recognize that the development of technology and the growth of research in the field would follow a pattern that begins with technological problems, proceeds through financial aspects, continues with institutional issues, and culminates with societal effects. This results in a body of literature that takes on a pyramidal shape, with the largest and oldest literature focusing on technical issues, and the smallest and newest exploring societal impacts.

It is obvious why this pattern occurs. In the first instance the development of the technology must proceed to a point where it is capable of performing desired operations reliably. Until this occurs, no implementation can occur, not even as a demonstration. Even after researchers succeed in producing a reliable technology, the diffusion of the technology is still a dream.

At this stage, the cost of the technology is frequently a major impediment to its diffusion. We are all familiar with the proliferation of videocassette recorders in the homes of ordinary people. When the first Sony Betamax machines hit the market at around $1000 each, only the most dedicated videophiles and electronophiles rushed out to buy one. Nowadays, at under $200 per VCR, and with hundreds of titles available for viewing, the videocassette recorder is a common household fixture. Similarly, early GISs were exceedingly costly. Moreover, because of their high price tag and the small number of adopters, there was little incentive for programmers to develop the capabilities of GISs to perform new tasks, and, therefore, to increase their utility for intended users. As researchers identified cost as a crucial factor in the slow early diffusion of GISs, continued advances in

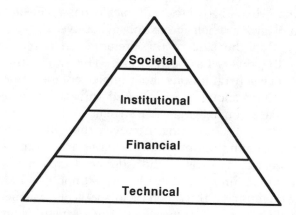

Pyramid of research on technology.

the technology itself addressed the double problem of high cost and limited capabilities.

As the cost of GISs declined and their capabilities increased, the diffusion of GISs has gained momentum. In fact, a question raised at the 1993 Conference of the Urban and Regional Information Systems Association wondered where organizations would get the money needed to purchase GISs in the future. It is a mistake to assume that tight budgets will automatically slow the diffusion of GISs dramatically, largely because of the declining prices and increased capability of the technology. The technology is becoming a better and better bargain. We need only look at the diffusion of personal computers in the late 1980s and early 1990s to see that even when average personal incomes are steady or in decline, lower costs can make possible the purchase of useful technology even in tight economic times. For example, in early 1989, the price of a Macintosh SE computer was approximately $2,500; 2 years later, its successor, the somewhat more powerful Mac Classic, cost about $1,000. Declining costs and increased affordability have been key in the diffusion of computer technologies and have made it possible for diffusion to continue in spite of tightening budgets. We assume that a continuation of this trend in GISs will render the same pattern of diffusion for this technology as well.

At this point in the diffusion of GIS technology, researches began to raise and discuss institutional issues in the literature on GISs. We discuss many of these issues in this book. Now, the GIS community is beginning to discuss societal issues surrounding the implementation of GISs. A workshop held in November 1993 provided one of the first major discussions in North America of GISs and society. This discussion is expected to stimulate further research in this area. The November 1993 issue of The Professional Geographer, "Open Forum: 'Automated Geography' in 1993," was an important addition to this topic (Lord, 1993).

This discussion is warranted by the growing diffusion of GISs worldwide. Still, in spite of the increased diffusion of GISs, scholars such as Goodchild and Getis (1991) suggest that users are only beginning to tap the full capabilities of such systems, with most of them using the technology primarily as a way to store data and information and few using GISs to perform spatial analysis. In effect, the capabilities of GISs have now outstripped the needs of the average user. This discrepancy between the capabilities of GISs and the needs of the average user, we suggest, is partly a function of institutional limitations, including such things as standard operating procedures. A secondary goal of this book is to pave the way for more complete use of the full range of GIS tasks by average users by providing some rudimentary information about geographical analysis and its value.

This concern brings to the fore a question that members of the GIS community frequently ask one another: "Are geographic information systems inherently different from any other type of information system?" The answer is yes and no. "No" in the sense that all information systems enable the user to collect, store, analyze, and output data according to the needs of the user. And if Goodchild and Getis are correct in their characterization of the ways in which GISs are used, then we can infer that for most users it *is* just another information system, with the exception that it is capable of using input and providing output in map form.

On the other hand, when implementation of a GIS goes beyond general information systems and the average user implements fully the geographic capabilities of the technology, then the answer is unequivocally and emphatically "Yes." This type of implementation will require not only information about how to operate a particular setup of hardware and software but also knowledge about the existence, range, and value of spatial analytic techniques that the GIS is designed to perform. These techniques grow out of theoretical and applied research in fields such as geography, landscape architecture, surveying, and others and include such operations as siting decisions, point pattern analysis, and network analysis.

This book is intended for use by a wide audience. We hope to appeal to the growing concern among members of the GIS community for more information and understanding about institutional issues related to the implementation of GISs. However, we also hope to appeal to the audience of organizations and people who have implemented, are implementing, plan to implement, or may be considering implementing a GIS system but wonder what a GIS can do for them that a nongeographic information system cannot. For both audiences, the theoretical foundation combined with examples is designed to promote a greater understanding of general principles related to the implementation of a GIS, which in turn is designed to improve the chances of successful implementation.

As authors, we believe we bring a distinctive perspective to concerns about managing GISs. Nancy Obermeyer has completed programs in political science and public administration, and has worked in state government in the areas of planning, budgeting, energy, and transportation before completing graduate programs in geography. Similarly, Jeffrey Pinto's studies are in the area of business and organizations. Both of us bring preexisting expertise in organization theory to our study of GISs, and we both developed our interest and expertise in GIS implementation as participants in research efforts by the National Center for Geographic Information and Analysis. We hope that our perspective will provide useful insights for anyone with an interest in the implementation of GISs.

REFERENCES

Aronoff, S. 1989. *Geographic Information Systems: A Management Perspective* (Ottawa, Ontario: WDL).

Goodchild, M. F., & Getis, A. 1991. *Introduction to spatial analysis*. Notes accompanying a workshop of the same name held in conjunction with the annual Geographic Information System/Land Information System Conference, Atlanta, GA.

Huxhold, W. E. 1991. *An Introduction to Urban Geographic Information Systems* (New York: Oxford University Press).

Lord, J. D. 1993, November. Open forum: "Automated geography" in 1993. *Professional Geographer, 45*(4), 431–460.

Ostrom, V. 1986. In F. X. Kaufmann, G. Majone, & V. Ostrom (Eds.), *Guidance, Control and Evaluation in the Public Sector* (Berlin: Walter de Gruyter).

Peuquet, D., & Marble, D. 1990. *Introductory Readings in Geographic Information Systems* (London: Taylor & Francis).

Tobler, W. R. 1976. Analytical cartography. *American Carographer, 3*(1) 21–31.

❖

Acknowledgments

One thing that we have both discovered about "writing" is that the noun should never be confused with the verb. That is, the idea of writing a book on the management of GISs was, and remains, a worthwhile endeavor, however, as we can both attest, the actual daily effort that goes into the writing process has led us both to another valuable insight: books are not written in a vacuum. Their successful completion depends upon a number of behind-the-scenes people whose impact, although they were not engaged in the actual writing, is felt no less keenly on this project. Special thanks are due to Bob Aangeenbrug (University of South Florida), Ron Abler (Association of American Geographers), Kate Beard (University of Maine), Will Craig (University of Minnesota), Andrew Frank (Technical University of Vienna), Harlan Onsrud (University of Maine), Mike Goodchild (University of California—Santa Barbara), Bill Huxhold (University of Wisconsin—Milwaukee), and Ian Masser (Sheffield University), all of whom are a tremendous source of ideas and inspiration for our work in this area and have influenced the way we think about many of the subjects we cover in this book. The National Center for Geographic Information and Analysis has also been a great resource for us. The Center as a whole, headquartered at the University of California at Santa Barbara, has made it possible for both of us to meet with members of the GIS community to discuss a variety of issues that have made their way into this book. Peter Wissoker and the editorial staff at The Guilford Press provided just the right kinds of moral and logistical support to enable us to bring this project to a successful conclusion.

Contents

O N E

❖

Introduction

The purpose of this chapter is to offer an introduction to the challenges that exist in managing within the field of GISs. These systems have evolved rapidly into a technology that can have a tremendous impact on the activities and productivity of a public agency or private firm. Because many of the issues related to the management of a GIS are still in their infancy, this chapter—and, indeed, this book—is intended to create a basis, both theoretical and practical, for the effective development and use of an organization's GIS. In this chapter, we intend to lay a foundation for the development of some of the many issues that concern managers in both public and private sectors who have a need for geographic information in their operations. In addition, we will present an overview of the book, discussing the various topics to be explored and their relevance for GIS managers.

THE NEED FOR A MANAGEMENT FOCUS

The GIS is becoming an increasingly common tool in organizations within both the public and private sectors. Increasing capabilities, decreasing costs, and enhanced user-friendliness have all contributed to recent gains in the rate of adoption and widespread use of GIS technology. However, managers charged with the task of creating an effective GIS for their organizations often find that they encounter a two-pronged problem: mastery of the GIS technology itself as well as understanding how to manage the effective introduction and use of geographic information. It is our position that the lion's share of material written on GISs to date has focused almost entirely on the area of explaining and teaching the GIS technology. This approach is

1

hardly surprising, given that the GIS technology is still young. Indeed, it is extremely important that a common body of knowledge about the technology be developed and shared among both researchers and users of these systems; furthermore, without that effort, there would be no need for information about managing GISs. This technical and introductory approach to GIS has resulted in some important writing, as evidenced by the work of individuals such as Robert Aangeenbrug (1991), Roger Tomlinson (1985), Michael Goodchild (1991), William Huxhold (1991), and others.

Now that GISs have been in the hands of users for several years, it seems appropriate to investigate the perceived satisfaction with the systems by their users and the problems that have arisen. Unfortunately, to date, anecdotal evidence provides a mixed response to the question: How satisfied are you with your system? While many users are highly satisfied and have made tremendous use of the various applications available with GISs, others have not been as enthusiastic in their evaluations. Among the comments we frequently hear are statements such as, "Nobody wants to take the time to learn the system," "Because the system is in our department, no one else wants to use it," "We never had any incentive to get it up and running, so there it sits," and "The policy makers never ask for any of the information we could provide, so we rarely use it."

If the technology works, where is the problem? Why is it often difficult to create the enthusiasm and commitment necessary to implement and use GISs successfully? The responses listed above are just a few of the comments that point to the larger problems that hinder gaining acceptance of any new innovation: the *will* and *skill* needed to manage its effective use. By *will*, we mean the desire to be proactive in seeking appropriate uses for a GIS as well as performing advocacy duty to infuse others with the same sense of enthusiasm for this innovation—in effect, to serve as a champion for one of these systems within an organization. Research and personal experience both demonstrate that where there is a will to invest in an information system and search out applications for its use throughout an organization, it is much more likely to be used and perceived in positive terms.

In addition to possessing a *will* to use GISs in organizational operations, many innovative technologies fail because of a lack of managerial *skills* on the part of key organizational members. These individuals are often in the position to manage the successful introduction and transition to new technologies but, for a variety of reasons, are unable to do so. This problem highlights a central thesis of this book: The success or failure of new technologies such as GISs is often directly related to the success or failure of an organization's management of the innovation.

We believe that the successful management of a GIS requires more than an in-depth knowledge of GIS technology. A GIS manager's mastery of the technical aspects of a GIS and spatial (or geographical) analysis is very important to an organization's eventual effective, efficient, and appropriate use of geographic information. Equally important for GIS success, however, is an understanding of some of the key organizational and behavioral dynamics that can determine whether a GIS will be used and, if so, how effectively it will be employed. In other words, we suggest that the quest for enhanced organizational productivity through introducing a GIS remains a managerial challenge every bit as much as it is a technical challenge.

Thanks to a growing body of research and writing, technical mastery of GISs has and will continue to become easier. As the technical expertise of many GIS managers and GIS teams has grown, we are beginning to see the focus of their concern shifting to a concomitant set of managerial problems that must be addressed. While these managerial concerns (such as successfully implementing the systems and coordinating activities across departments) have always been in place, in an organization's rush to embrace and learn a new technology it is understandable why new systems are initially given short shrift. However, the time to begin addressing these issues is at hand.

Research in a variety of fields as diverse as the study of new product introduction, project management, and the implementation of management information systems (MISs) has demonstrated the importance of managerial skills as a basis for system success. To illustrate, a study of project management conducted in the 1980s by Baker, Murphy, and Fisher (1983) sought to determine which factors are most important to project success. Among the wide range of factors that they considered were technical mastery of various project elements, as well as such behavioral and managerial factors as leadership style, motivation, maintaining effective working relationships with other project stakeholders, and team building. Their research found that these behavioral factors were far more important for project success than the more technical details. This finding has been replicated in a number of other studies conducted on the implementation of new information system technologies. These findings have reinforced the importance of understanding the managerial roles and organizational dynamics needed to facilitate system acceptance and effective use.

In this chapter, we will lay the groundwork for the discussion that takes place in the rest of the book, as we address some of the variety of managerial issues that confront organizations attempting to achieve the maximum effectiveness from their newly acquired GIS. We will begin by defining and describing GISs. We will then discuss in an overview fashion some of the topics that will be developed in greater detail in each of the subsequent chapters.

WHAT IS A GEOGRAPHIC INFORMATION SYSTEM?

This may seem to be a very rudimentary question, but it is nonetheless crucial to the reader's understanding of this book. Chances are, you have heard the initials "GIS" at least once elsewhere (otherwise, you probably would not have picked up this book in the first place). A short definition of a "geographic information system" is a computerized system for the collection, storage, manipulation (analysis), and output of information that is spatially referenced. The problem with this short definition is that it ignores the fact that other types of information systems are also spatially referenced—the inclusion of street addresses or zip codes instantly makes them so.

What sets GISs apart from all other types of information systems are their reliance on spatial referencing as their organizing framework and their ability to perform geographic analysis. For many people, the acronym GIS is synonymous with maps. While some GIS professionals may be disappointed by such an understanding, it is actually a good starting point. There are other members of the GIS community who acknowledge that a map is itself a geographic information system—a noncomputerized, manual system, but a GIS nonetheless.

How can this be? A map is a means by which we *collect* geographic data. For example, a typical city map contains data points locating specific points of interest (the county courthouse, the regional shopping mall). It contains lines showing roads, highways, and interstates. A city map contains polygons, showing the boundaries of city parks, historic districts, and so forth; if the map is on a large enough scale, polygons may also show the footprints of buildings. As we look at a map, we do not see it as a meaningless mishmash of data points, lines, and polygons. Rather, our eye sees the map as a well-organized image of a city, with buildings, streets, and neighborhood parks.

In creating any map, the cartographer pays attention to specific rules and procedures. In so doing, he or she produces a stored image that those who are familiar with cartographic conventions can understand and use. Rand McNally has built a successful company on this premise. Their road atlases are probably among the most commonly found items in the average person's automobile. Without a doubt, the map is a means of *storing* spatial data.

Are these data in a form that allows one to *analyze* them? Yes. The fact that millions of people use Rand McNally's Road Atlas to take them from sea to shining sea in North America is a testimonial to the ease of analysis that a map represents. Similarly, the Michelin road maps have helped people find their way around the world for decades. Way-finding is a basic type of geographic analysis. There are others.

For example, we use maps to make siting decisions. For the average, noncommercial user, a prime siting decision is where to live. In coming to a new city to take a job, a person and his or her spouse might first start by locating their places of employment. This hypothetical couple might identify the location of specific schools (if they have school-age children). They might locate shopping districts, parks and playgrounds, museums, or any other "must have" neighbors, as well as bus routes to these sites. On the other hand, they might also identify potentially noxious neighbors, such as polluting industries, rundown neighborhoods, prisons, red-light districts, and so on. Using this information, the couple can then identify neighborhoods that meet their basic specifications. Most people will need to do another round of analysis at this point, identifying which neighborhoods contain affordable housing and which are too pricey.

Similarly, commercial interests frequently make decisions about where to locate their outlets. For example, in recent years, Toys "R" Us has expanded into two new markets,—Bangor, Maine, and Terre Haute, Indiana. Surprisingly, Bangor, whose population is around 30,000, opened its store a year before the opening of the Toys "R" Us in Terre Haute, whose population is about double that of Bangor. On the face of it, this seems like an odd decision; however, anyone with geographical training and familiarity with both places can rather easily infer the reason for this corporate decision.

Both Terre Haute and Bangor are regional centers. Terre Haute attracts retail business from the south and west (into Illinois), whereas Bangor attracts retail business from the west and the north (into Canada). Terre Haute has competition from other cities about an hour away, including Indianapolis, a thriving city of approximately 1 million people that has attracted many corporate headquarters and management centers in recent years. By contrast, Bangor is the largest city for hundreds of miles to the west and north, and thus has no real competition. Moreover, Bangor Mall, a neighbor of Toys "R" Us that predates the outlet by 10 or more years, already had a record for attracting consumers from throughout northern Maine, as well as from the Canadian provinces of Quebec and New Brunswick. The Canadians must drive a minimum of 2 ^{1}/$_2$ hours to reach Bangor; many drive much farther. One reason that Canadians are willing to drive such long distances is the large price advantage of shopping south of their border. A quick perusal of the Bangor Mall parking lot supports the claim by Victor Konrad, former director of the University of Maine's Canadian–American Center, that roughly one-third of the Mall's income is from Canadian consumers.

When we have these bits of information, we can understand that while the city of Bangor is smaller than Terre Haute, its retail sector has a very large market area—geographically much larger than that of Terre Haute. Furthermore, as a retail center, Bangor has almost no competition, whereas Terre Haute is a mere 1-hour drive via Interstate 70 from Indianapolis with its many large shopping centers. In fact, so easy is the drive that Terre Hautans readily make it on a whim. In this light, the Toys "R" Us decision makes perfect sense.

Without getting bogged down in detail, let us mention that understanding this decision is made possible by the knowledge of several basic principles of geography: central-place theory and its discussion of lower and higher order goods, and spatial interaction models, with their ideas of complementarity, transferability, and intervening opportunities. It is by understanding these principles that we begin to get a handle on what the GIS can do.

Maps may be used to produce a number of different types of analysis, including way-finding and siting decisions. The results of the analysis of maps is often a new map as *output*. This new map will have combined the data available from the preexisting map in such a way as to produce new information to answer a question. The new map may show where to target your house-hunting expeditions. The output may also, however, be a report. For example, the Toys "R" Us corporate office probably ended up with a planning document that identified 1989 as the year for Bangor's outlet, and 1990 as Terre Haute's year. As you can see, it is easy to think of maps as a noncomputerized GIS.

And this brings us to a basic truth about GIS. For those who have never used GISs and who have not studied geography or another spatial science, the novelty of these systems and sciences lies in the "new" things they can do. In fact, geographers and other spatial scientists recognize that a GIS represents a new and improved way to do what we and our predecessors have always done. No more, no less. In the past, we performed these analyses the old-fashioned way, with map, ruler, pen and paper. Later, the development and diffusion of the calculator and the computer helped us to avoid hours of laborious calculations. The GIS is a breakthrough in much that same vein, making spatial analysis easier still.

Ironically, in spite of the growing range of spatial analytic capabilities of GISs, some scholars in the field (Goodchild & Getis, 1991) make the point that most GIS users do not even begin to tap into the power of these systems. Instead, most users rely on their GISs as a means of storing and updating data and of producing maps. While GISs can perform these tasks handily, they are capable of doing a good deal more than this. The types of analyses described above—and many other analyses as well—are well within the current capacities of today's GIS.

Why, then is the GIS underused in this way? One answer lies in the minimal training in geography prevalent in North America, particularly in the United States. While most of us have had one or two years of geography in grammar school, we learned more about state capitals than about the whys of geography: Why is that thing where it is? What effects will its location have on things around it? These are the general questions that geographers ask, and that GIS can help to answer. These questions form the basis for the principles of geography.

A sound understanding of basic principles of geography will assist in the conceptualization of the full range of existing and potential uses to which a GIS can be put. If we were to make one suggestion to those who are unfamiliar with the raw potential of the computerized GIS, it would be to take the time to study geography. Enroll in an introductory physical or human geography course as a means to learn more about the capabilities of geographical analysis. Take an advanced geography course in the area of your work (e.g., urban geography, economic geography, climatology, hydrology) as a way to gain new insight into the spatial dimensions of your field.

The above discussion is intended to achieve two objectives. First, we wish to highlight the importance of effective management practices for enhanced GIS operations. Mastery of the technical aspects of the GIS is not enough to create an environment for successful GIS applications. An important complement to the technical details of geographic information is understanding some of the behavioral and organizational issues that can facilitate or impede the effective use of a GIS. This emphasis on management of a GIS—to date virtually ignored in many of the major works in the field—is the primary focus of this book.

Our second objective from the previous discussion is to place GISs and GIS professionals within the context of an emerging profession—one that has faced and will continue to face a series of technical and managerial challenges. These challenges must be faced head-on if GIS professionals are to demonstrate the efficacy of their systems in public and private sector organizations and, thus, encourage the subsequent adoption of this innovative technology by other organizations. Many of the challenges that are inherent in the emergence of a new profession are also addressed in this work, including personnel issues and certification, policy conflicts and the role of GISs, and the challenges of using GISs within a democratic society. It is with these two foci in mind that the book has been written and organized.

ORGANIZATION OF THE BOOK

Chapter One

In this chapter we offer the basic philosophy behind which this book was written. We define GISs and give examples of the types of analysis they can perform. In addition, we argue for the need for a managerially focused approach to the successful operation of GISs.

Chapter Two

This chapter introduces some of the basic issues of organization and behavioral theory within the context of a discussion of the most efficient and effective implementation of GISs. As we discuss basic facets of organization theory, we also make the case for the importance of understanding organizations as a means to enhance the prospects for successful implementation of GISs. The whole issue of implementing

GISs is discussed in some detail, drawing upon pertinent literature and findings from other fields as we trace the processes by which organizations successfully introduce and implement new innovations, including information systems.

Chapter Three

It is important to place a GIS within the correct context as part of an organization's overall information system. This chapter lays a foundation for some of the important uses of information systems within public and private organizations (usually termed "Management Information Systems" [MIS]). We examine some of the basic roles that an MIS performs for managers in their decision-making processes. Further, we introduce geographic information into the overall scheme of MISs to demonstrate its specific contributions to enhanced managerial decision-making.

Chapter Four

GISs represent a departure from typical organizational policy development tools because of their explicitly geographic component. As the use of GISs becomes more common, their potential will become more readily apparent. One of the keys to unlocking the potential of GISs is a better understanding of geography among GIS users. This chapter is intended to shed light on the importance of geographic and cartographic principles for various GIS applications using examples from public policy analysis and cartography It includes a three-point approach to using a GIS in order to maximize its effectiveness to an organization.

Chapter Five

In this chapter we discuss problems of lack of familiarity with geography and GISs as organizational impediments to the adoption of GISs. We suggest that the deeply rooted bureaucracies that exist within public and private organizations are often the root cause of difficulties with GIS acceptance and use. Understanding the basic problems of bureaucracies and their attitudes toward many innovations offers the key to using the organization's bureaucracy to your advantage in attempting to adopt a GIS.

ONE

Chapter Six

Given the costs of GIS, adopting the technology is an important decision, especially in the current era of public budget belt-tightening. In this chapter we examine the economic justification of adopting a GIS. In addition to the traditional discussion of benefit–cost analysis, we also explore basic principles of public goods economics that complicate the problem of justifying a GIS. These principles include problems of pricing public goods and the substitution of private goods for public goods.

Chapter Seven

As thsee systems become more widespread, organizations find it increasingly useful to share information across organizational boundaries. These boundaries occur between organizations that can mutually benefit from sharing geographic information as well as within a single organization where different departments readily offer each other access to their information. In this chapter we discuss the impediments to such sharing and suggest methods to promote greater inter- and intraagency cooperation.

Chapter Eight

During the 1980s, strategic planning gained widespread acceptance as a key management activity in both the public and private sectors. Strategic planning (or management) represents an improvement over comprehensive planning in that it assumes change and uncertainty and proposes to manage them through routine monitoring of both its external and internal environments coupled with an ability to respond quickly and appropriately to change. Not surprisingly, strategic planning is information-dependent. GISs represent a useful tool for information management and analysis in strategic planning. These issues are explored in this chapter, as are some of the basic steps or aspects of the planning process in understanding types of information that are important in environmental scanning.

Chapter Nine

A crucial component of the successful performance of one of these systems is the competency of an organization's personnel. The nature of a GIS requires a balance of expertise in three areas: geography, computers, and the specific functional area of the organization (e.g., transportation, human services). In this chapter we attempt to enhance managerial success in balancing these areas of expertise by promoting an awareness of some of the related issues, including problems of technocracy and the development of credentialing in GISs.

Chapter Ten

One of the claims made by advocates of GISs is that these information systems can help to minimize conflicts by providing more and better (more accurate) information. This claim overlooks an important source of conflict: the underlying value differences represented by conflicting parties. In this chapter we explore the role of GISs in the public policy sphere and suggest that GISs will tend initially to increase rather than decrease conflict, since geographic information and analyses made possible by GISs will be used selectively by conflicting parties to support their positions. However, this conflict can be viewed as a positive feature of a democracy, representing the open expression of differences of opinion that must be explored fully as a precondition for acceptable public policy resolutions.

Chapter Eleven

In this chapter we discuss the problems and challenges of implementing a GIS in a democracy. The collection of information required by implementing a GIS raises some important issues related to societal control as a result of the possession of large amounts of information that a GIS makes possible. We examine these problems in an attempt to heighten the awareness of GIS managers to their public responsibility. Further, we offer an overview of some of the important legal ramifications of this enhanced public access to information. Where do the rights to privacy conflict with the fundamentals of freedom of information?

CONCLUSIONS

This book represents our efforts to offer a challenge to the professional GIS community in acknowledging the importance of effectively managing their systems. The technology underlying GISs has been in place for some time and continues to be upgraded with enhanced features and increased user-friendliness. If we are to promote the successful performance of GIS technologies in our organizations, it is time to consider some of the managerial roadblocks that have prevented more widespread organizational acceptance and use of these innovations. It is our purpose to suggest some of the first steps in this direction, by reviewing some of the more popular behavioral and organizational aspects of managing GISs.

Implementing a GIS:
Theories and Practice

A necessary part of the evolution of ever more complex information systems has been the introduction of these systems into organizations to aid in their day-to-day operations. One of the long-standing difficulties with the development of enhanced information system capabilities within organizations has been the problem of gaining widespread acceptance and use of these systems. This implementation problem has continued to inhibit the successful implementation of information system technologies throughout their history. It is our contention that the primary problems that underlie most implementation concerns are usually organizational, rather than technical. That is to say, in contrast to technical problems accompanying the introduction of new innovation or information system technologies (which are usually quickly surmounted), organizational problems of implementation refer to the human aspects that can inhibit or limit the potential acceptance and use of new system innovations. A considerable body of research evidence has shown that paying attention to the behavioral or human factors within the organization can greatly facilitate the likelihood of acceptance and use of new technological innovations.

Numerous studies have addressed the adoption and diffusion of technological innovations in other fields (e.g., Rogers, 1962; Schultz & Slevin, 1975; Downs & Mohr, 1976) However, few formal social science investigations of geographic information technologies have been reported (Wellar, 1988a, 1988b; Wiggins & French, 1991; Onsrud & Pinto, 1992). It is important to note that the terms "diffusion" and "implementation" are used interchangeably within this section. Historically, the concepts of diffusion and implementation were independently derived from different sources. ("Diffusion" usually referred to the acceptance and use by some subset of the general population of scientific or technological innovations. The concept of "implementation" derived from the acceptance within organizations of new technical processes or models.) However, over the last several years, the two terms have come to engender the same idea. Therefore, for the purposes of this chapter, diffusion and implementation are both intended to refer to *the process through which an innovation is communicated through certain channels over time among the members of a social system* (Rogers, 1983).

An understanding of the diffusion process can help those who could benefit from an innovation, such as a new technology, to begin accruing those benefits earlier. By identifying crucial social factors and processes in the adoption, implementation, and utilization of a technology, we would expect to predict the decision-making responses of individuals, groups, and organizations more accurately, and, therefore, to accommodate or redirect these processes through prescriptive strategies. By identifying crucial human and technical factors within classes of potential users, diffusion studies also have the potential for directing the design efforts of system developers to those system characteristics and improvements most valued by end users.

In this section we will explore some of the important issues in the diffusion and implementation of new information system innovations, examining the historical roots of the implementation problem, some of the approaches to implementation that have been examined to date, and the nature of successful implementation. We will suggest, through the development of a framework model, that implementation success is a multifaceted issue, comprising a number of diverse but equally important criteria.

The concept of implementation in the context of organizations may be viewed as a change phenomenon or a process for creating organizational change. Initially, the problem of implementation was discussed in the context of frequently ineffectual attempts on the part of operations researchers and management scientists to generate enthusiasm for and use of a myriad of new organizational innovations intended for use by practicing managers. More recently, the problem of implementation has been represented as the frequent failure to create some degree of desired organizational change through the introduction of a new information system, program, or model. Indeed,

Schultz, Ginzberg, and Lucas (1983) define implementation success in terms of changed behavior on the part of organizational members. Implementation, then, in its most basic sense, has increasingly come to be defined as a new information system, program, or model that has been accepted by organizational personnel, the results of which system change the decision-making processes of the personnel (Schultz, Ginzberg, & Lucas, 1983; Schultz & Slevin, 1975).

Literature in the area of implementation, though significantly increasing in volume in recent years, remains largely unfocused. That is to say, many writers in the field have discussed implementation in a general sense, paying little attention to the type of implementation effort being performed, perhaps in the belief that there exists little or no real difference between various forms of organizational implementation (information systems, projects, OR/MS [operations research/ management science] models, etc.). Bean, Neal, Radnor, and Tansik (1975) discussed implementation of OR/MS as seemingly interchangeable systems and projects. Further, Harvey (1970) developed a set of critical success factors in the implementation process without regard to distinguishing between types of implementation.

Early work on implementation tended to focus on the important actors in the implementation process rather than on the type of implementation being considered. For example, Churchman and Schainblatt (1965) discussed cognitive styles and the need for mutual understanding between the researcher and the user as a way of improving the chances for implementation success. Schultz and Slevin (1975) argued the need for formal feedback channels to encourage constant communication between researcher and user, in the belief that a key focus of implementation research should be on the organizational actors.

The topic of implementation has also received a great deal of attention in the public policy and public agency literature. Pressman and Wildavsky (1973) coauthored one of the earliest works on implementation, discussing the implementation problem from both theoretical and practical perspectives and using the public policy forum as a basis for their work. Further, Majone and Wildavsky (1978) discussed the concept of implementation as a controlling and interacting variable, distinguishing between activities necessary for public policy enactment and those related to the implementation of previously developed policies. While much of the public policy work on implementation has largely remained unintegrated with the rest of the field of organizational research, the work of Wildavsky and others demonstrates the strong connection that implementation has with both public and private sector problems.

DEFINITION OF IMPLEMENTATION SUCCESS

Because implementation has been previously defined as a change process that is accepted by organizational members, a definition of successful implementation must incorporate these components. There are several definitions of implementation success, many of which share some common underlying properties. Early definitions of implementation success were usually predicated simply on use or the acceptance of the new innovation by its intended user (Mitroff, 1975; Lucas, 1975). The implementation process was generally seen as a bridge between system developer and user, a bridge that once having been crossed, was regarded as successful.

Other definitions of implementation success have been more complex and thorough in their interpretation of the factors relating to this success. Arnoff (1971), delineating conditions for implementation success in operations research models, stated that "unless our models result in decision rules and/or procedures, (a) which are implemented within the managerial decision making process, (b) which really work, and (c) which lead to cost effective benefits, our efforts are not successful" (142).

Schultz and Slevin (1979) offered an important contribution to the discussion of implementation success by supplementing these decision rules with a behavioral perspective of implementation. Their approach distinguishes three criteria by which any implementation should be judged: *technical validity*, *organizational validity*, and *organizational effectiveness*. Technical validity and organizational validity are seen as necessary but not sufficient conditions for the implementation process to result in a positive change in organizational effectiveness.

Technical Validity

The first criterion for implementation success, *technical validity*, is the belief that the system to be implemented works—that is, it is a correct and logical solution to the perceived problem. Obviously, should a new information system package be fraught with difficulties and other system bugs, it will not be used, and, hence, should rightly be regarded a failure. The Ashton–Tate Company lost a great deal of money and prestige in attempting to ship a new release of its flagship package DBase IV too early, before significant system problems had been debugged. Companies attempting to implement the new database management system would have been unsuccessful owing to the lack of technical validity in the package.

Organizational Validity

The second factor, *organizational validity*, is a measure of the congruence between the organization and the system to be implemented. As such, it may be considered as the belief that the GIS is correct for its users and that it will be accepted and used by the organizational members for whom it is intended. The key advance in system implementation theory that is suggested by the notion of organizational validity is the importance of including acceptance and actual use of the system in any equation of implementation success.

A classic manifestation of organizational validity is the "not invented here" (NIH) syndrome held by many corporations. NIH refers to the parochial attitude that a product or idea is of no value unless it is created in-house. Simply put, the belief is that an externally derived idea will not be used by the company because it came from outside sources. Companies laboring under NIH frequently report serious difficulties implementing new systems or other innovations because, while being technically correct, the systems will not be used by the client groups for whom they are intended.

Effectiveness of an Organization

The bottom line that any organization seeks through the development and adoption of a new system or other innovation is some measure of improvement, either in productivity, efficiency, or environmental effectiveness. Early researchers recognized the importance of demonstrating some payoff from adopting a new information system. Based on the work of Schultz and Slevin (1975) and Schultz and Henry (1981), it was further determined that new definitions of implementation success need to be based on the perception of implementation as a process of organizational change. The basis for this view is the argument that it is not sufficient to regard an implementation as successful merely because the intervention of a new system causes organizational personnel to change their decision-making processes—in effect, to make use of the model. Successful implementation requires that the results of the system lead to *improved* decision making. As a consequence of this definition, a new information system, such as a GIS, that is in place but is not improving decision making (i.e., leading to enhanced organizational effectiveness) would not be considered successfully implemented.

IMPLEMENTATION MODELS

Geographic information systems are a somewhat unique technological innovation in their current state of development in that they require a combination of both centralized and decentralized processes for their effective implementation through most classes of potential users. The classical implementation conceptual model presumes a centralized structure with a technological innovation originating from some expert source. Under this model, the innovation development process for a community of potential users begins with recognition of a need or problem; moves through research, development, and commercialization of the innovation; continues through implementation and adoption of the innovation by users; and ends with the consequences of the innovation (Rogers, 1983). However, a GIS is a multipurpose tool offering advantages to different classes of users who disperse them at different rates (e.g., utilities vs. planning agencies vs. scientists vs. delivery services). Within each class, considerable adaptation or reinvention appears to occur before the operational characteristics and information product capabilities are perceived as beneficial across the class.

At the adopter level, decentralized implementation processes are often required in order to meet the differing database development needs of each organization, along with the needs of groups and individuals within each organization. Decentralized communication and technology-transfer processes occur among and within similar organizations that are all involved to varying degrees in adapting the innovation to their circumstances. While the implementation of GIS hardware and software generally follows the classical model, the implementation of data characteristics and data handling methods appropriate to the class of potential users (e.g. types of data, quality and accuracies of data, means of collecting, forms of storage, forms of system-generated products) probably can be explained best by a decentralized implementation model. Thus, in studying the implementation of geographic information technologies, researchers need to identify both centralized and decentralized implementation processes as well as significant reinvention processes (Rice & Rogers, 1980).

The above discussion suggests that a geographic information innovation, the implementation of which might be promoted or evaluated, could consist of a sole hardware/software combination, a broad range of commercially developed or in-house-developed geographic information-processing capabilities, a unique and useful data set or database method, a standard for data collection, and/or some other technological or institutional development. Thus, an innovation of concern might be any identified innovation that some members of a class of users have found beneficial and that is predicted to be adopted by a high percentage of the social class over time, with or without adaptations to the innovation. In this section, the geographic information innovation typically used for discussion and illustrative purposes consists of computer-based GISs (i.e., the broad class of systems, referred to as GIS, currently being widely introduced throughout industry and government).

GIS IMPLEMENTATION STUDIES

From an implementation research perspective, it is gratifying to note that a great deal of recent literature in the GIS field has focused on the process of implementation of technological innovations within end-user organizations. Conferences and agenda-setting groups have increasingly expanded the realm of GIS research beyond the original development- and applications-centered descriptive studies to include investigations centered on understanding issues of GIS acceptance and use (Onsrud & Pinto, 1991).

Much of the existing research on implementation of GIS technologies possesses similar properties regarding methodology and research design. Indeed, the vast majority of GIS implementation research consists of single-case studies in which practitioners and academic researchers report on the success or failure of their particular implementation effort within a local government, planning agency, or some other end-user site (Wentworth, 1989). From these case examples, conclusions are drawn that are presumed to be generalizable to the larger population of similar users. For example, Levinsohn's (1989) experience with the introduction of GISs led him to conclude that top management must be involved in major automation decisions. More recently, Antenucci et al. (1991) used case illustrations and examples to develop five types of implementation activities: concept, design, development, operation, and audit. As Table 2.1 demonstrates, they were able to make some preliminary distinctions between successful and unsuccessful system introductions on the basis of a host of managerial issues, including planning, staffing, funding, and so forth.

In studying the implementation, use, and assessment of geographic information innovations, there exists a broad range of quantitative and qualitative research methods to choose from. Among these are mathematical modeling, controlled experiments, quasi experiments, surveys, longitudinal studies, field studies, archival and secondary research, futures research and forecasting, content analysis, case studies, focus groups, and interpretive and critical approaches that have developed in response to shortcomings of the positivist methods (Williams, Rice, & Rogers, 1988). No one method is most appropriate for studying a broad or complex research problem. Each method or combination of methods has advantages and disadvantages as well as different assumptions, biases, and degrees of usefulness (Williams, Rice, & Rogers, 1988). In fact, a recent approach of research scholars has been to emphasize the use of several research methods in combination in order to accommodate the weaknesses of each method with the strengths of others (Kaplan & Duchon, 1988).

TABLE 2.1. Elements of GIS Project Success and Failure

Activity	Characteristics of GIS Projects	
	Success	Failure
Planning	Rigorous	"Run and gun" style
Requirements	Focused	Diffused
Appraisal of effort	Realistic	Unrealistic
Staffing	Dedicated, motivated, high continuity	High turnover
Funding	Adequate	Inadequate, conjectural
Time estimates	Thoughtful	Rushed or prolonged
Expectations	Balanced	Exaggerated

Note. From Antenucci, Brown, Croswell, and Kevany (1991). Copyright 1991 by Chapman & Hall. Reprinted by permission.

While current GIS case studies have shed valuable light on some of the important steps in and characteristics of the GIS implementation process, they are hampered by several drawbacks. Case studies and other qualitative forms of social science research have long been criticized for their limitations regarding generalizability to the larger population and lack of sampling controls (Cook & Campbell, 1979; Piore, 1979; Bonoma, 1985). Further, retrospective reporting of successful or unsuccessful implementation efforts is often subject to considerable information loss and bias, particularly when substantial time has elapsed since the implementation effort occurred. Finally, some of the published case-study research involves the reporting of the implementation process by a single individual (usually the manager responsible for the implementation). Obviously, there is strong temptation for these individuals to report implementation experiences in the best possible light, even if they are somewhat inaccurate. In a recent article Wellar (1988a) articulated many of the problems existing in the current research paradigm focusing on qualitative case methodologies.

Some, but not all, of the shortcomings of current GIS case-study approaches may be overcome by using more logical and rigorous case research methods. In recent management information system (MIS) literature, Benbasat, Goldstein, and Mead (1987) provided some suggestions on how to conduct and evaluate information system case-study research and provided some examples of research areas particularly well suited to case approaches. Lee (1989) and Onsrud, Pinto, and Azad (1992) extended that work to present a scientific methodology with which to conduct case studies of MISs and GISs, respectively. They argued that through an analysis of scientific method—especially the four requirements that a scientific theory must satisfy (i.e., making controlled observations, making controlled deductions, allowing for replicability, and allowing for generalizability)—researchers may identify the point at which scientific rigor is achieved in case-study research, beyond which further rigor, particularly at the expense of professional relevance, is questionable. Lee also argued that a qualitative study of a single case can possess more analytic rigor than a statistical study using sophisticated numerical analysis tools. Onsrud, Pinto, and Azad (1992) suggested, however, that the reverse is often more typical. Regardless, important goals for researchers investigating geographic information implementation issues should include becoming familiar with the benefits and drawbacks of the wide range of research methods available, selecting research methods that are appropriate to acquiring the knowledge sought, and selecting a series or suite of methods that is designed to balance the weaknesses of each individual method with the strengths of others.

Although the eventual public and private investment in improved geographic information-handling capabilities is estimated in billions of dollars, few studies have attempted to correlate the usefulness of the technological innovation with adoption, use, and abandonment or to evaluate the efficacy of the technology in social terms. The assumption prevailing among GIS professionals seems to be that because the technology is being adopted it must be valuable and useful. Yet little information is available, other than in anecdotal form, to support or disprove this assumption. As a result, the study of the implementation of GISs represents an important issue that has not been adequately supported by research efforts to date.

ASSESSMENT OF IMPLEMENTATION SUCCESS

Utilization

One of the more intriguing and continually debated issues in research on the implementation of innovations concerns the assessment of technological implementation success. In its simplest sense, success implies the degree to which the implementation effort was perceived to be successful. Such a definition, however, begs the larger questions of determining exactly what constitutes implementation success and when and how success or failure should be measured. Past implementation studies in other fields frequently presumed that upon confirmation of the acquisition of a technological capability, the innovation was successfully implemented. That is to say, adoption success and implementation success were considered synonymous. This logical error continued even after organizational theorists argued that the failure of implementation studies to produce consistent findings was due largely to the failure to identify stages of implementation in those studies (Zaltman & Duncan, 1977; Greer, 1981).

In evaluating the transfer of geographic information innovations through a social system, attempts to isolate crucial adoption factors and processes with their temporal dependencies (e.g., when they occur) and correlations with organizational attributes (e.g., what organizational factors influence them) will be valuable but will not be wholly adequate. For reasonable expectations of consistency and generalization of results, it will be necessary to expand studies to address the full process of adoption (i.e., acquisition, initial implementation, and use of the innovation by the organization). Those organizations within a class having already acquired the geographic information innovation under consideration should be evaluated in regard to postacquisition implementation, extent of utilization in the organization (e.g., Goodchild & Rizzo, 1987), levels in the organizational structure of that use, forms of decision making utilizing the innovation, factors and processes leading to rejections of the previously embraced innovation, and abandonment patterns over time.

One interesting hypothesis long espoused informally among GIS professionals and noted also by technology-transfer researchers in other fields is that within an organization the influence that various individuals and groups have on the acquisition of a technological capability may be only marginally connected to actual staff decisions to use the technology (Greer, 1981). For instance, the authors can cite an example in which operational geographic information-processing capabilities were delivered through a contract for an unrelated purpose to the staff of an organization who had little say in the acquisition but who now heavily utilize the system. In yet other examples, the staff appear to have been involved from the very beginning in acquiring the technology, and yet the technology appears to be largely underutilized by staff in the organization. Because GIS implementation is still at the early adopter stage for most classes of potential users, the field is lacking in attempts to study the use of the innovation within and across classes. However, one lesson suggested from past experience in other fields is that after base information is acquired on individual and organizational adopter characteristics, including correlations among the characteristics and with attributes of the innovation, expansion of the field of enquiry to utilization phases may be more fruitful in understanding the implementation process than first trying to directly probe deeper into adoption questions.

Because utility in decision making has proved so difficult to measure, Ives, Olsen, and Baroudi (1983) have developed a method for measuring user satisfaction with information systems: they argue that user satisfaction may serve as an appropriate surrogate for utility in decision making (see also Raymond, 1987; Igbaria & Nachman, 1990). This approach is also worthy of consideration in the arena.

Impact Assessment

A technological innovation such as GISs is of little consequence until put into general use. Although the effects of technology on both a social system and on the values of that system are extremely important,

Rogers (1983) notes that the social consequences of innovations have received very little attention from implementation researchers and change agents. Again, a prevailing assumption has been that if the customs of a social class are altered through general embracement of an innovation, the social consequences, as judged by the adopters, must be beneficial; otherwise, the innovation would not have been embraced. This reasoning in many instances is false, as evidenced in the literature by the numerous examples showing that adoptions of innovations have had highly adverse consequences for an industry or social system as a whole (Niehoff, 1966). The assumption that adoption equals success, in fact, results in a pro-innovation bias. Clearly the findings of social system impact assessments should be held out for consideration by those who have not yet adopted innovations, in order to remove the bias and the underlying fallible assumptions. However, it is difficult to measure and evaluate the social and economic responses of classes of users to an innovation. These conceptual and methodological difficulties are at least partially responsible not only for the current lack of research and of generalized findings on social consequences but also for the pro-innovation bias.

Identifying the times at which the effects of innovations should be assessed is difficult. If assessment occurs too early in the implementation process for the class of users, an innovation may not yet have had time to be adapted fully to users' needs. If assessment occurs too late, it may simply memorialize the fact that large numbers of adopters made a wise decision or a poor decision to invest in the particular innovation. Ideally, assessment should occur early enough in the implementation process to offer practical guidance to system designers and later adopters and yet not so early as to result in a gross underestimate of the innovation's effectiveness and value. In yet another sense, innovations such as GISs are continually evolving, and, hence, there also exists a need to assess them in relation to a social class each time significant technological capabilities relevant to the needs of that class are developed.

One of the obvious problems in attempting to assess the effect of an innovation on a particular organization or across a broad class of users is the difficulty in separating out the effects of the innovation from the effects of other changes in the institution that have occurred contemporaneously with the innovation's implementation and use. Associated with this is the widely acknowledged problem that many of the benefits of information systems are indirect and are, therefore, difficult to measure or estimate quantitatively (Dickinson & Calkins, 1988; Money, Tromp, & Wegner, 1988).

Although a technological innovation may be attractive through traditional cost–benefit analysis, it may have adverse effects on the overall effectiveness of an organization. The reverse is also true. These conclusions suggest a need to consider which individuals or units gain or lose in an organization with respect to the ability of each individual or unit to contribute, the quality of work, and the control of financial and other corporate resources (Greer, 1981; George & McKeown, 1985). Difficulties remain, however, in assessing these factors as well as in determining whether and to what extent such factors are likely to contribute or distract from the long-term efficacy of an organization or to the long-term efficacy of an entire class of users. Because of this, there is a need to develop alternative methods and strategies for assessing the effectiveness of geographic information innovations at many different levels, such as individual investment in the technology, performance of an entire organization or business, or overall performance of an industry, discipline, or other broad class of users. (Bie, 1984; Williams, Rice, & Rogers, 1988).

Most difficult of all is assessing the societal consequences of an innovation. The societal effects of GISs are potentially very great. These systems have a promising future in helping various segments of society to address some of the more pressing social problems. Such systems offer efficiencies and capabilities that were previously unavailable and are already being used by individuals, government agencies, private businesses, and a range of organizations to deal with resource management and environmental problems ranging from site specific problems to global scale issues. However, even if one were able to confirm that the cumulative effects of decisions made with the aid of GIS have helped increase the overall quality of life and the efficiency of resource production within a social system while decreasing immediate and long-term effects of development on the environment—all of which are far from trivial to assess—adverse and potentially severe social consequences are also likely to arise from implementation of the technology.

One area of concern repeatedly addressed in the geographic information literature are potential shifts in the access rights of citizens to information (Roitman, 1988; Onsrud, 1989; Archer & Croswell, 1989; Epstein, 1990). Many government agencies have established operational land-information systems and are making decisions based on analysis of data within those systems that directly affect the daily lives of citizens (i.e., taxing, permitting, service delivery, zoning, districting, and similar decisions). Will such systems increase citizen access to information and promote equal access, or will these systems—particularly in light of recent local government initiatives in the United States to alter state open-records laws in the cause of cost recovery or user-fee strategies—create substantial differentials in people's ability to access publicly held information? In addition, the likely effects of geographic information technologies on laws and policies relating to work product protection (i.e., copyright, patent, trade secret, etc.), rights to privacy, confidentiality, liability, and security have not been widely studied nor have strategies for lessening adverse consequences been fully explored.

Determining what constitutes a beneficial versus a detrimental consequence is a value-laden judgment. Chrisman (1987a) suggests that equity is the primary principle around which GISs should be developed, so that all affected by use of information in the system will be treated fairly. Rogers (1983) implies that in distributing the consequences of innovations, a strategy should be developed and employed that will decrease, or at least not increase, the magnitude of socioeconomic gaps among members of the social system affected by an implementation program. Geographic information researchers should consider developing goals, policies, and strategies that will promote increased equity in the distribution of the beneficial consequences of geographic information technologies.

CONTENT AND PROCESS MODELS OF IMPLEMENTATION

One of the serious problems with past research into the implementation of innovations has been the use of either content or process models as the sole investigative heuristic. A content approach to implementation analysis focuses on determining those specific environmental, organizational, and interpersonal factors that can facilitate or inhibit the implementation process (Leonard-Barton, 1987). Process approaches, on the other hand, strive to analyze the key steps or decisions in understanding how innovations are diffused. While each method is useful, neither offers a complete picture. A thorough approach should identify both the key decision factors in adopting geographic information technologies and the processes by which the implementation occurs.

Implementation of GISs: Content Models

Under a content model approach, data are typically collected from a limited number of case studies in an attempt to identify those implementation model variables that are significant to the adoption process for the particular class of potential users being considered. Past implementation research suggests to us that in the GIS environment we should be particularly cognizant of potential crucial factors in the following areas (Rogers, 1983; Onsrud, Calkins, & Obermeyer 1989; Raghavan & Chandf, 1989; Huxhold, 1991; Croswell, 1989):

- Visibility of benefits
- Complexity in learning or using the innovation
- "Trialability" of the innovation
- Compatibility with existing values, past experiences, and needs
- Relative advantage of the innovation over the product, process, or idea that it supersedes
- Social norms
- Existence of formal and informal communication channels
- Appropriate balance between mass media and interpersonal communication channels
- The extent and accessibility of vendors of an innovative technology
- Opportunity for information sharing among colleagues
- The presence or absence of champions, reinventors, and opinion leaders in organization or in the professional community
- How adoption decisions are made within the organization or peer group
- The extent of reinvention necessary to adapt to local circumstances
- The extent of consensus on methods and standards
- Consequences of adopting innovations
- The likelihood of unanticipated repercussions from adoption of an innovation
- Memory of past failures
- The presence of backups if something goes wrong

The above list is not exhaustive. For instance, one potential factor affecting adoption that could be included in the above list is the economic advantage provided by the innovation. In most technology adoption studies to date, however, the economic value of an innovation appears to play a relatively minor role in the decision to actually embrace the technology. Any institution considering investment in an innovation must first cross the threshold of having enough slack in its resources to be able to make some initial investment in it. However, presuming the slack is available, factors other than immediate economic advantage typically are shown to be far more crucial in the actual decision to adopt. Ad hoc observations of adoptions of GISs, for instance, suggest that numerous institutions and organizations are investing in GIS capabilities even though traditional cost–benefit analyses indicate that the investment will never pay for itself over the life of the software–hardware system being purchased. In other instances, although cost–benefit analysis strongly supports an investment in GIS capabilities, the organization has been loathe to incorporate the capability. The differences in the adoption decisions probably can be explained largely by isolating the crucial implementation factors. What at first appears to be an irrational economic decision is converted to a rational decision when the crucial factors in the implementation process are taken into account.

The lists of content factors developed by these and other researchers often vary in degree of comprehensiveness, from broad general outlines (consequences of adopting the innovation) to specific points for consideration (presence of a champion). In spite of their diversity, it is possible to discern some general factors that have been found to be crucial to new-system implementation success. These factors were originally posited within the context of new-project implementation; however, their general nature makes it possible to draw parallel lessons for the implementation of GISs as well (Schultz, Slevin, & Pinto, 1987). The factors are as follows.

1. *Clearly defined goals* (including the general philosophy or mission of the organization, as well as a commitment to those goals on the part of key organizational members earmarked to use the system). All parties within the organization affected by the implementation of a GIS need to be aware of exactly what tasks they and the new system are expected to perform.

2. *Sufficient resource allocation.* Resources in the form of money, trained personnel, logistics, and so forth, are available to support the newly installed system.

3. *Top-management support.* Top management within an organization has made its support for the project known to all concerned parties.

4. *Implementation schedules.* A well-detailed plan for new system implementation, including training time, has been prepared and disseminated to all concerned parties.

5. *Competent technical support.* The manager and support personnel for the system installation have the necessary experience and technical competency to ensure a smooth transition to the new system.

6. *Adequate communication channels.* Sufficient information is available on the system's objectives, status, changes, organizational coordination, user's needs, and so forth. Further, formal lines of communication have been established between the implementation team, the system's intended users, and the rest of the organization.

7. *Feedback capabilities.* All parties concerned with the system can review its implementation status and make suggestions and corrections through formal feedback channels or review meetings.

8. *Responsiveness to clients.* Any of the system's ultimate intended users are clients. All potential users of the newly installed system are consulted and are kept up to date on the system's status. Further, they will continue to be assisted after the system has been successfully implemented.

As the above list demonstrates, typical implementation efforts often follow relatively similar patterns as far as the most important factors are concerned. Further, as has been mentioned earlier in this chapter, it also becomes quickly apparent that many, if not most, of these factors are more managerial than technical in nature. Implementation theorists and researchers have known for some time that problems with the diffusion and adoption of new technologies are often based on human issues rather than on technical difficulties or concerns. This is not to suggest that a system does not need to be technically adequate in order to be accepted. However, as Schultz and Slevin's (1975) development of the ideas of organizational validity and acceptance demonstrates, the battles for successful information system implementation are usually won or lost, not in resolving all technical issues relative to the GIS, but in appealing to and attempting to address organizational members' concerns.

As already mentioned, the problem with the content model approach is that—while it offers important information for managing a system's implementation—it is essentially a static representation of the implementation effort. In other words, content models do not mirror the importance of the process by which a new system is implemented. Most definitions of implementation have included as part of their description phrases such as "the process of organizational change" or a "change process." Consequently, implementation models need also to reflect the dynamic nature of new-system diffusion. In other words, the content model approach has value and provides insights but should be used as only one of the components in a comprehensive implementation model.

Implementation of GISs: Process Models

Unlike content models of implementation, which are aimed at identifying the factors that are the key determinants of innovation acceptance and use, process models are concerned with determining the key phases in the adoption process. One model suggests that there are two main subphases in the innovation process:

1. *Initiation.* The organization becomes aware of the innovation and decides to adopt it.
2. *Implementation.* An organization engages in the activities necessary to put the innovation into practice and incorporate it into existing and developing operations.

These terms, "initiation" and "implementation," have been redefined by Schultz, Slevin, and Pinto (1987) as the "strategy" and "tactics" of an implementation effort. Their argument suggests that one simple but effective way to view the implementation process is as a distinction between planning activities and action-oriented efforts. That is to say, planning activities (termed "strategy") are related to the early planning phase of the implementation process. They represent either the conceptualization of the new-system implementation or its planning and control. A second set of factors (referred to as "tactics") is concerned with the actual process, or the action of the implementation, rather than its planning.

The essence of Schultz, Slevin, and Pinto's argument is that conceptualizing system implementation as a two-stage process has further implications for system performance. Figure 2.1 shows the breakdown of strategy and tactics by low or high score depending upon the level to which these issues were addressed in the implementation process. For example, a high score on strategy would imply that the strategy was well developed and effective. This value could either be assessed in a subjective (or in an intuitive) manner or more

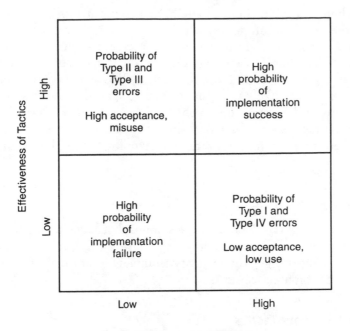

FIGURE 2.1. Strategy/tactics effectiveness matrix. From Schultz, Slevin, and Pinto (1987). Copyright 1987 by Randy L. Schultz, Dennis P. Slevin, and Jeffrey K. Pinto. Reprinted by permission.

objectively, using some surrogate measures of the implementation process (such as initial organizational member buy-in, schedule and budget adherence, and so forth). A manager could, for example, determine that the strategy was deficient, based on past experience that he or she had with other, successfully implemented systems.

We can speculate on the likely outcomes for new-system implementation efforts, given the assessment of their strategic and tactical performance. Figure 2.1 illustrates the four possible combinations of evaluated performance of strategy and tactics. The terms "high" and "low" imply strategic and tactical *quality*—that is, the effectiveness of operations performed under the two clusters.

Four types of errors may occur in the implementation process. The first two error types were originally developed in the context of the statistical testing of hypotheses. The last two error types have been suggested as the result of research on implementation. Type I error occurs when an action should have been taken but was not. Type II error is the taking of an action when none should have been taken. Type III error means taking the wrong action (solving the wrong problem). Type IV error occurs when a solution is discovered but is not used by an organization (Schultz & Slevin, 1975). Each of these error types is more or less likely to occur depending upon the mix of strategy effectiveness and tactics effectiveness. Knowing the interaction of strategy and tactics and their probable effect on project success and potential error type is important in understanding the implementation process.

High Strategy, High Tactics

Quadrant 1 in Figure 2.1 holds those systems that have been effective in carrying out both strategy and tactics during the implementation process. Not surprisingly, the majority of implementation efforts under this classification are successful.

Low Strategy, Low Tactics

The reciprocal of the first quadrant is in the third, where both strategic and tactical functions were inadequately performed. We would expect implementations falling in this quadrant to have a high likelihood of failure.

Low Strategy, High Tactics

While the probable results of new-system implementation are intuitively obvious in quadrants 1 and 3, the results for efforts that fall in quadrants 2 and 4 are not. Quadrant 2 represents a situation in which initial strategy functions were insufficient, but subsequent tactical activities were highly effective. Some of the expected consequences for system implementations falling into this category would be an increased likelihood of Type II and Type III errors. Type II error would occur in a situation in which initial strategy was ineffective, inaccurate, or poorly developed. However, in spite of initial planning inadequacies, goals and schedules were operational-ized during the tactical stage of the implementation. The results of Type II error could include implementing a poorly conceived or unnecessary system that has received no initial buy-in from potential users and either may not be needed or will not be used.

Type III error may also be a consequence of low strategy effectiveness and high tactical quality. Type III error has been defined as solving the wrong problem, or taking the wrong action. In this scenario, a need has been identified, or a new system is desired; but owing to a badly performed strategic sequence, the wrong problem was isolated and the subsequently implemented GIS has little value in that it does not address the right target. In this case, the tactics to develop and implement a new system are again well-conceived, although initial planning and problem recognition were poorly done. The result would be high acceptance of the system, as well as its general misuse, owing to action being taken to implement a system when none may have been warranted.

High Strategy, Low Tactics

In quadrant 4 lie cases where strategy was effectively developed but subsequent tactics are rated low, or ineffective. One would expect implementation efforts classified in this quadrant to show a likelihood of Type I and Type IV errors (not taking an action when it has been determined that action is needed or simply not using the new system). To illustrate, consider a situation in which strategic actions have been well performed, thus suggesting the installation of a GIS. Type I error would occur when little action is subsequently taken and the tactical activities are so inadequate that the new system is not implemented.

Finally, Type IV error would occur following an effective strategy that has correctly identified the need for a GIS for an organization; but after poor tactical operationalization, the system was not used by the clients within the organization for whom it was intended. In other words, Type IV error is the result of low client acceptance. As discussed earlier, the reasons for lack of acceptance are numerous, but most often they revolve around a failure to match the system to the needs of the organization and its specific personnel.

Process models of innovation have been useful for identifying the important steps in gaining acceptance and use of new innovations. However, process models, by themselves, are also afflicted with some difficulties (Srinivasan & Davis, 1987). First, as mentioned earlier, most process approaches in the past have been highly qualitative in nature. Second, they do not attempt to determine who the key players within an organization are at each step in the adoption decision. Third, these models may vary or fluctuate to a significant degree

depending upon the type of innovation that the organization is trying
to adopt—in other words, the models may not be sufficiently
generalizable. Fourth, it is often the case that the desire for innovation
may exist within an organization at a particular node, whereas the
overall organization is not particularly innovative. For example,
certain nodes of an organization, such as engineering or software
development, may seek to adopt a GIS, while the overall organization
is reluctant to pursue computer-information system innovations.

CONCLUSIONS

For managers attempting to better understand the management of
geographic systems within their organizations, a basic knowledge of
organization theory and human behavior is essential. In this chapter
we have argued that many, if not most, of the problems associated with
managing the introduction and use of a new information system are
people problems rather than problems associated with technical
difficulties. Implementation theory and research have, for years,
known that the most prevalent implementation problems (such as lack
of acceptance and use) as are the result of poor development of an
organization's human assets. Consequently, any discussion of the
process by which a GIS is introduced and managed within an agency
or organization must be predicated on developing a greater understanding of the organization as a social system.

❖

The Role of Geographic Information within an Organization's MIS

The purpose of this chapter is to offer an introduction to the development and use of an integrated management information system (MIS) within an organization. In order to understand the implications and use of geographic information, it is necessary to place the GIS within the context of a larger, fully integrated system that provides managers with relevant information for performing their duties. We will show that the purpose behind an MIS is to aid in managerial decision making by providing an organization's members with data and other forms of information that are comprehensive, comprehensible, and of immediate use. As a result, in this chapter we will develop an overall picture of the importance of MISs for modern organizations, the duties of management and the information needs across various management levels, the nature of the classical decision-making process and the ways in which an integrated MIS can affect managerial decision making, the manner in which information is gathered and processed, and the specific role that geographic information plays within the context of this larger framework.

The society within which we exist and operate has become—and will continue to become—increasingly complex and fast paced. Within the private sector, competition has reached the international level in a number of industries. As a result of these external pressures, the cycle time for new product innovation has been forced to decrease rapidly in an effort to speed up time to market in order to meet consumer needs.

The public sector is equally affected by the faster and faster pace of our society. Local, county, and federal governmental agencies are being called on to take an increasingly proactive role in the management of infrastructure, land use, natural resource development, surveying, and a host of other activities related to the more efficient management of our urban and rural environments and natural resources.

Information has become a valuable and often expensive resource in today's society. The rapid rise in the creation and expansion of MIS departments within organizations lends credence to the importance that is attached to providing managers with timely and useful information to enable them to better perform their duties through more effective decision making. In order to make clear the role of information within the operations of organizations, it is important to define exactly what is meant by the term. *Information* is data that has been converted, or operationalized, into a meaningful and useful context. Once such a context has been agreed upon, the information is of considerable value to specific organizational members, who use this information in an effort to arrive at better (more effective or efficient) decisions.

Owing to the increased need for precise and useful information, one of the more recent phenomena in the field of organization theory has been the rapid rise of information system development and use. An information system (IS) refers to a system of people, resources, and procedures that collects, transforms, and distributes information to relevant organization members. For an IS to be effective, it must supply managers with information that is rapid, comprehensive, and accurate. Further, it is important to emphasize that the most important aspect of information provided by a system is its usefulness to its end users—that is, managers. While there may actually be many methods for collecting and disseminating information, for the purposes of this chapter we will concentrate on the activities of computer-based ISs.

AN OVERVIEW OF MANAGEMENT

It has been stated above that MISs were developed as a tool to enable managers to better perform their jobs. Such an observation, however, begs the larger question of the role that managers are expected to play for organizational success; in other words, what is it that managers do? In order, then, to gain a sense of the effect of ISs on the process of management, it is important to devote some attention to a discussion of exactly what constitutes management. In other words, what are some of the specific duties and activities that correspond to the role of a manager within a public or private organization? Literally thousands of books have been written over the last century on the process of management, how it works, what the specific duties are, and how to improve it. Yet, of all the works on the study of management over this period of time, one of the most influential is *General and Industrial Administration* by Henri Fayol (1916). In this book Fayol outlined his views on the proper management of organizations and of the people in them. He presented five primary roles of management: planning, organizing, supervising, staffing, and controlling, which although over 75 years old, have formed the basis of almost all subsequent work in the field of management. Each of the five primary tasks are defined in the following paragraphs.

Planning

The role of *planning* requires that the manager develop a set of goals and objectives and create both long-range and short-term plans for achieving these goals. Long-range plans are often broad, general outlines of where a company or a specific department wants to be in 5, 10, or even 20 years. Short-term goals are established to address and focus attention on specific targets that the organization seeks. These targets are seen as contributing and complementary to the organization's efforts to achieve its long-term objectives. For example, at General Electric in the early 1980s, Jack Welsh, the then-new chief executive officer of the organization, formulated his famous "one, or two, or out" rule. In other words, each operating division within General Electric would, within 2 years, either become number one or number two in its specific product industry or it would be sold. With this long-term objective as their backdrop, operating managers within each of GE's strategic business units were required to formulate short-term plans for gaining a commanding share of their individual markets.

Organizing

Organizing refers to the methods by which managers organize, or make sense of, the work environment. The standard methods used to organize include the development of an organizational structure and operating rules and procedures. To illustrate: As an organizing function, a manager may choose to change the nature of the reporting structure within his or her specific department, either increasing or decreasing the number of levels of management between the workers and himself or herself. Another example of the use of organizing through standard operating procedures could be the establishment of a rule that all purchases or other expenses in excess of $1,000 must be approved by the department manager.

Supervising

Above all else, managers need to recognize that their primary responsibility is that of human resource management. In other words, managers are essentially involved in the *supervising* role. Their success or failure hinges on their ability to develop, motivate, and lead their subordinates. As effective leaders and motivators, managers are required to provide their employees with both the opportunity and the means to be productive. Further, within their supervisory function, managers are sometimes called upon to act in a guidance mode with their subordinates, offering counseling and support for those who need it.

Staffing

The process of *staffing* involves the selection and professional development of organization personnel. In essence, it refers to the manager's responsibility to ensure that the right person, with the right training, occupies the right job. Staffing activities can actually be quite varied, from performing interviewing and hiring duties to providing job and skill training opportunities for organization personnel.

Controlling

Controlling refers to a manager's duty to monitor the activities of his or her subordinates in order to ensure that all activities are performed effectively and, in cases where deviations from plans are noted, to provide the necessary corrections. As a result of the monitoring process, it may be noted that either employee performance or initial plans will need to be modified. That is to say, if the manager observes that subordinates are performing to their maximum and yet are unable to achieve the targeted objectives, it would then become necessary to modify the initial projections in order to bring the objectives more into line with reality. On the other hand, if employees are not performing up to their potential, the manager may have to provide additional training or even correction and discipline, if appropriate. Control is most often found in the form of feedback, whereby a manager receives a report on each subordinate's job performance and, on the basis of this information, determines whether or not some form of correction is needed.

The common factor underlying each of the five duties of management is the need to make timely, informed, and accurate decisions. In order to make the most efficacious decisions, managers need to receive the types and quantity of information that will enable them to best perform their jobs. It is with this purpose in mind that organizations have developed and introduced a variety of ISs into their operations.

MANAGEMENT INFORMATION SYSTEMS

As stated earlier, it is the primary purpose of MISs to provide managers with information that is complete, accurate, and timely and that will enable them to make decisions that are more efficient and effective. As noted by Hutchinson and Sawyer (1992), an MIS is created to satisfy a manager's need for information "that is more summarized and relevant to the specific decisions that need to be made than the information normally produced in an organization and is available soon enough to be of value in the decision-making process" (471–472).

Information systems serve the needs of managers in two ways. First, they provide a sense-making function in that they assist management in understanding the complex nature of the relationship between the organization and its environment. By making available needed information of a readily usable nature, managers are able to make more informed and, arguably, better decisions. A good IS aids in gathering data and processing internally useful *intelligence information* as well as externally disseminated *public information* (Schermerhorn, 1989). Within an organization intelligence information is the basis upon which key decision makers chart long-term objectives. Public information is derived from the environment and allows an organization to engage in a wide variety of public activities, including image building, advertising, political support, and so forth.

The second way in which an IS serves the needs of managers is through timeliness. Obviously, information that arrives late or incomplete is of almost no value. In order for managers to reap the advantages of an IS, the system must materially influence the *way* in which managers arrive at decisions as well as the *type* of decisions they make. Further, managers need to be aware in their own minds that these new, information-assisted decisions are in some sense superior to the old method, either through time savings or enhanced effectiveness of the decisions themselves. It is also important to note that there is little in common with the types of decisions made at different levels within an organization. As a result, an MIS must provide a variety of differential pieces of information so that the information can be accessed and can assist in supporting the decisions made by managers at different levels in an organization.

INFORMATION NEEDS ACROSS ORGANIZATION LEVELS

The first step in designing an IS is to determine what sorts of information management needs. This task is more difficult than it may at first appear, particularly as one moves upward within an organization's hierarchy. Top managers are often required to perform in a capacity that is characterized by greater ambiguity and that requires more generalized knowledge rather than specific functional expertise. Table 3.1 is intended to illustrate this point and presents a summary of some of the different types of activities performed at various organization levels. Further, it suggests the types of information that would be most useful to and appropriate for managers across these different organization levels. As you can see, first-level managers usually perform tasks that consist of implementing operational plans as developed by higher-level management. For example, a first-level manager's duties may include scheduling production runs, assigning resources across various tasks, and transacting day-to-day business activities. Consequently, the types of information that first-level managers need is usually tied directly to the specific tasks they are called upon to supervise. Defect reports, exception reports, and adherence to budgets

TABLE 3.1 Information Needs at Different Management Levels

Level	Activities	Information needs
Top management	Strategy formulation: the establishment of long-term objectives and plans, making strategic decisions	Wide ranging: many sources are required, both internal and external to the organization; there are external opportunities and threats and internal strengths and weaknesses
Middle management	Formulation of plans for achieving strategic objectives: making a specific variety of operational decisions including resource allocation, employee evaluation, short-term goal setting	Mostly internal: includes a combination of general and specific information requirements
First-level management	Performance of well-defined tasks: making short-term decisions, transacting day-to-day business	Diagnostic: designed to enable correction of deviations from specific schedules and budgets; best information is measurable

and schedules are examples of some of the concrete, specific types of information that are useful to first-level managers and that can be operationalized. Ideally, any diagnostic information is valuable that can help these individuals perform their duties more efficiently.

Middle-level supervisors also have a variety of duties, oftentimes of a more general and ambiguous nature than those assigned to first-level managers. Middle-level supervisors are usually called upon to find methods for implementing higher-level strategies. As a result, they are tasked with the need to formulate operational plans and objectives that will allow for the successful implementation of business strategies. Further, they make operational decisions in support of these plans. For example, a middle-level manager who has been charged with increasing productivity in a series of midwestern plants may act to fulfill that requirement by reallocating human and financial resources to the midwestern region. As you can see, the type of information that a middle-level manager may require is more general and wide-ranging than what would be needed by a first-level supervisor. In addition to simple production reports, the middle-level supervisor in this example would also need financial and profitability data, as well as information relating to manpower and to budgetary slack.

Finally, top management operates in a very different manner from managers at other organization levels. Top managers engage in such activities as formulating long-term goals, making strategic decisions regarding corporate direction, determining products to be developed and produced, and securing a variety of scarce resources on which the company depends for survival. In order to most effectively engage in new strategy formulation and goal setting, top managers require a wealth of information that is not of interest to managers at lower levels. For example, many of the information needs of top management are external; that is, these needs require that an organization's IS analyze and provide data on general trends in the marketplace, on changes in governmental and economic policies, on consumer patterns and tendencies, and so forth. These types of information are in direct contrast to those needed by lower-level managers. First-level managers are often provided with concrete, tangible information that enables them to compare actual progress to production or output plans and, where appropriate, make necessary corrections. On the other hand, top management, which is engaged in a series of more ambiguous activities, requires a wealth of additional information from the external environment in order to chart the most effective courses for the organization in the future.

MANAGERIAL DECISION MAKING

Up until now, we have made the point that MISs enable managers to make better decisions by providing more complete information. However, an important point that needs to be considered is the process by which this information is normally incorporated into the decision-making process. In other words, how *do* managers use information to make decisions? What role does information play in the decision-making process that warrants such an investment in information system technology on the part of many organizations? Once we understand the part played by information in decision making, we can begin to see that information systems have become an integral part of the process of effective decision making. As such, it then becomes possible to suggest: (1) the *stages* in the decision-making process at which information becomes important, and (2) the *types* of information that are most useful at these various points.

A phenomenal amount of research has examined the process by

which managers make decisions, in an effort to prescribe more efficient and effective methods. It should come as no surprise, however, that many managers make decisions in highly idiosyncratic ways. Some individuals engage in large-scale data exploration, while others make gut-feeling decisions following limited investigation (or even in seeming contradiction to the preponderance of existing information). However, when decision making is approached systematically, we can see the existence of a number of important steps. These steps often typify, in a general sense, the approach to decision making taken by most individuals. The specific steps are (1) problem recognition and diagnosis, (2) solution generation, (3) alternative evaluation and selection, (4) solution implementation, and (5) feedback (see Figure 3.1).

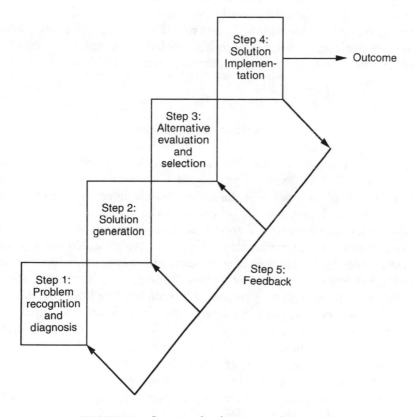

FIGURE 3.1. Steps in the decision-making process.

Problem Recognition and Diagnosis

Problem recognition and diagnosis refers to the acknowledgment that a problem exists. A problem simply means a difference between a planned state of events and the actual state. For example, a planned state would be a situation in which a county planning board has forecast the need for repaving 500 miles of road following winter and has budgeted money to cover that amount. However, if in the spring it was found that 750 miles of roads needed to be repaved, the county would clearly have underestimated the costs of road repair. A problem would then exist.

In many cases, it may be relatively easy to identify (diagnose) the source of the problem and begin remedial steps. However, other problems do not lend themselves to such simple diagnosis. For example, a car that failed to start one morning could be signaling a number of different problems that would have to be checked and eliminated one at a time before the owner could be reasonably sure that the correct problem had been identified. As a result, the first step in the process of decision making is recognizing that a problem exists and engaging in a systematic search to make the correct diagnosis of the cause of the problem.

Solution Generation

The second step in the standard decision-making model involves the *systematic search for a series of possible solutions to the recognized problem.* For some problems, the potential list of possible alternative solutions is seemingly endless, while in other cases the solution may be quite bounded, either because the correct solution is obvious or because the organization has a limited amount of resources to engage in a search for alternatives. In the example of the county highway commission, the set of alternatives is rather narrow. Brainstorming may generate a solution set that would include: asking for additional funding from the county board of supervisors, putting off repaving 250 miles of road until the next round of budget meetings, stretching the budget and material resources to do a superficial job of repaving the entire 750 miles, and reappropriating money from another source within the budget. It is important, however, to note that at this stage solution alternatives are only being generated, not evaluated. Prematurely evaluating the various choices often has the effect of inhibiting some individuals from taking part in the brainstorming process. The goal of this decision-making step is to develop the widest possible set of alternatives, not to make value judgments on any one suggested option.

Alternative Evaluation and Selection

After each alternative has been carefully explored in terms of strengths, weaknesses, possible ramifications, and long-term effects (if any), the next step is to *select the alternative that best meets the manager's or organization's objectives*. The logical choice is the alternative that maximizes a decision maker's initially developed set of objectives. In other words, the best choice is often that which offers the most benefits for the least cost. However, an additional constraint is whether or not the best alternative is in conflict with other goals and objectives of an organization. As a result, assuming a complementarity between organization goals and the optimal alternative, the decision maker will then make the appropriate selection.

Solution Implementation

Once a decision has been reached concerning the best course of action, it is necessary to act on that choice. *Implementing the solution simply means following through with a specific action or series of actions in order to solve the problem.* At this point a manager must be cognizant of the potential for acceptance of or resistance to the proposed solution on the part of other individuals affected by the decision. A decision alternative may arguably be the optimal choice and still be passed over if the likelihood of its acceptance is low. Recent history highlights the fact that the Coca-Cola Company introduced its New Coke brand as a replacement for Coca-Cola Classic for what the company's management believed were correct business reasons. However, consumer backlash was so severe that the company was forced to cancel its strategic move and return to maintaining its flagship product.

Feedback

The final component of the decision-making process is *some form of feedback channel that allows management to follow up on an implemented decision and evaluate its utility.* Management assesses the degree to which the originally agreed upon objectives are being achieved as a result of the implemented decision and, if necessary, modifies the decision or takes some other form of corrective action. Further, as Figure 3.2 demonstrates, it is necessary that feedback opportunities be available at each stage of the decision-making process. If, for example, the decision makers discover that the decision is incorrect, the cause of the problem may be the result of incorrect problem diagnosis or alternative selection. Therefore, they may need to cycle back through the process and reevaluate their initial assumptions or set of alternatives in order to make a different selection.

The benefits of using an MIS as a critical component of the decision-making process are obvious. First, because an IS is often a storage site for a variety of internally and externally derived pieces of data, it is invaluable in helping the decision makers to generate a selection of alternative solutions to any problem. Once a set of problem and decision parameters are specified, data can be retrieved from the system that is appropriate to the problem at hand. Further, as selection criteria are developed, an IS gives managers the power to

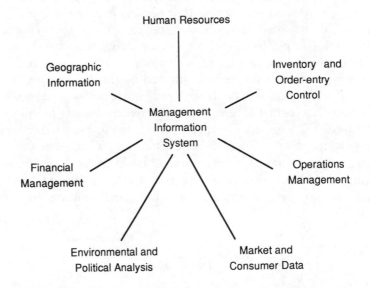

FIGURE 3.2. Geographic information as part of an overall management information system.

generate a set of optimal solutions that are specifically intended to maximize the alternative selection criteria. In effect, because data are stored in the IS, an organization can use its computer to process the possible solutions and then, relative to the selection criteria generated, determine the optimal solution that maximizes organization goals within the constraint of the likelihood of decision-implementation acceptance.

Finally, a computerized IS offers the best possible sort of feedback system for the decision-making process. Once solutions have been implemented and the results determined, that information forms a basis for future problem-solving activities. To illustrate: If the organization is faced with a problem that is similar to one that occurred at some point in the past, it is possible to retrieve all important information pertaining to that problem from the IS. This information would include the nature and symptoms of the original problem, alternatives generated, solution selected and implemented, and the results of that choice. Consequently, an MIS can serve as an integral part of an organization's decision-making process, particularly when data entry is well maintained and continually upgraded.

MAJOR COMPONENTS OF AN MIS

Although an MIS can be used as a highly effective decision-making tool to help managers identify problems, generate alternatives, and follow up the results, it is important that we gain a grounding in some of the major components of ISs. Specifically, there are four basic elements or types of activities that go into the development and effective use of an MIS: (1) data gathering, (2) data entry, (3) data transformation, and (4) information utilization. These categories are generally consistent with the components of a GIS (discussed in Chapter One), including data collection, storage, analysis, and output.

Data Gathering

Data gathering refers to the gathering of any and all data that may be deemed pertinent to an organization's operations. A local community, for example, would want to gather such information as census and tax-roll data regarding its residents; infrastructure data concerning roads, bridges, and other public works; planning and zoning maps for commercial and industrial development; data on optimal delivery of emergency services (fire and police); and so forth. Some data may be readily accessible (e.g., census figures), and other information may be harder to gather (e.g., possible local toxic waste storage sites developed by private companies). Nevertheless, at the data gathering stage an organization needs to determine what data are relevant to their activities and develop a plan for gathering the relevant data.

Once an organization's management team has determined which types of information can aid their operations (a potentially lengthy activity), there has to be a concerted effort to collect data. At this point a number of organizations make the mistake of assigning too few individuals to data gathering. Later discussions with managers at these same organizations reveal the managers' frustration with the seeming lack of utility of the newly developed MIS. This lack of utility is not surprising. The greater the number of organization members actively seeking and collecting information, the greater the possibility that an organization's MIS will have a significant positive effect on organization operations. It is important to remember that any computerized system is limited by the amount of data it is able to retrieve.

Data Entry

Data entry consists of any activities necessary to take the raw data that were gathered during the first step and put it into a digital form that can then be entered into computer data banks. One expensive activity that is currently taking place with a number of local and county governments is the digitizing of maps into computer-accessible forms. The old paper maps that have, for hundreds of years, been a basic tool of local governments are being converted into digital formats for computer data entry. The entry process is often long and tedious, as computer specialists convert raw data into a computer-usable format.

Data Transformation

Data transformation refers to reconfiguration into useful forms of data stored on the computer, that is, forms that are useful to managers querying the IS. For example, a number of computer software packages and structured query languages (SQLs) have been developed to assist managers in recombining or restructuring data in order to provide the specific information needed. To illustrate: If a plant manager wishes to determine the relative cost level of production for one shift at his or her plant, simply accessing the raw data might be of only minimal value to him or her; on the other hand, if some software packages were available to transform that data into cost structures, ratio analyses, and trend projections, the information would be much more accessible to the manager and, consequently, more useful.

Information Utilization

Information utilization refers to the idea that information can be retrieved as needed by management personnel and used in making a wide variety of operational decisions. This final element in an MIS consists of the actions taken as a result of the information that has been provided to managers. In other words, information that is not relevant or useful is simply wasting a manager's time. On the other hand, the ultimate assessment of an IS utility is the effect it has on enhancing managerial decision making. If, therefore, the information that has been provided is used, it is often appropriate to judge the efficacy of an MIS by user satisfaction with it (Ives, Olson, & Baroudi, 1983). In fact, to an increasing degree over the last 10 years, assessments of an MIS's impact are often measured through the surrogate of user satisfaction (Doll & Torkzadeh, 1988; Galletta & Lederer, 1989; Igbaria & Nachman, 1990; DeLeon & McLean, 1992). Utilization often represents an appropriate bottom line determinant of MIS utility. Because information is an essential element for aiding in managerial decision making, an organization's nonuse of available information offers a serious indictment of the strength of and need for an MIS.

GEOGRAPHIC INFORMATION AND MISs

In the previous sections of this chapter, we have developed the role of management information in the decision-making process by addressing some of the most common types of management activity, looking at the decision-making process, and detailing the basic elements of an MIS. With this background, this section is intended to establish the basis for the relationship between an organization's overall MIS and the specific role of geographic information.

Here the question that we must ask is, Is a GIS any different from an MIS (or any other type of IS)? As noted earlier, both use a computerized system for collecting, storing, analyzing, and outputting information. Obviously, MISs and GISs have much in common. In fact, as already noted, many types of ISs—not only GISs—have the ability to store and retrieve information by location via use of a zip code, a full address, or some other spatial identifier.

What sets GISs apart from other types of ISs is their ability to catalog spatially referenced objects and their attributes within the context of a map. Even more dramatic is their ability to perform quantitative analyses based on geographical principles. Thus, the GIS is different from other ISs.

Within an organization, geographic information often represents a subset of the overall database. Because geographic information by nature is more specialized data, it is appropriate for and useful to a specific set of organizations, both public and private, that are involved in various activities for which geographic information is both relevant and important. In the private sector, some organizations that find geographic information useful are natural resource exploitation and development firms (mining, forestry, drilling and gas exploration), building and building-supply companies, farming and ranching, and so forth. Within the public sector, the majority of local, county, state, and federal agencies need a variety of geographic information, including resource conservation and wildlife protection, infrastructure development and repair, zoning, tax-roll updating, land management, surveying and mapping, census-data analysis, and land-use analysis.

It is apparent that for a number of organizations, access to and use of geographic information is not simply a luxury but a necessary and integral part of any IS that they develop. For these organizations, geographic information forms the core of their operational database. Any problems or opportunities that they seek to address must be done within the context of querying and acting upon geographic data. As a result, geographic information serves as an important element within the framework of the larger organizational information system set up to aid in management decision making (see Figure 3.2). If a GIS is analyzed in relation to the background of managerial decision making that we have explored in the chapter, it is clear that geographic information, like other forms of information in an organization's database, is useful in making optimal decisions for solving specific problems. Further, the more directly an organization is tied into land and natural resource usage, the more important geographic information becomes in relation to other data sets within the overall information system.

CONCLUSIONS

This chapter has sought to provide an organizational context for the creation as well as the rapid and widespread acceptance of MISs within organizations. An MIS is an important and useful tool that can help managers make better (more accurate and effective) decisions by providing them with a more complete source of information, with decision alternatives, and with possible ramifications of various choices made. Further, an MIS plays a key role throughout the various levels of organization management because it enables managers at all levels to access and make use of information that is specific to their operations. At higher management levels, MIS is useful in organization policy and strategy development because it represents a repository of a variety of internal and external information that allows policy makers to spot trends and make necessary strategic moves. For lower-level managers, an MIS can provide the kind of diagnostic and prescriptive information that empowers them to operate at peak efficiency.

Geographic information is rapidly gaining in popularity and will continue to be a much sought-after resource for a number of public and private organizations. In developing an integrated GIS, it is important that managers be aware of the natural complementarity between the geographic information that is often key to their operations and other elements in an organization's overall MIS. The role of an IS is to offer managers enhanced decision-making capabilities by providing ready access to information (through the computer) and, as a result of data transformation, putting this information into usable formats that can have a timely effect on operations. In this chapter we have sought to establish the conceptual and practical link between a GIS as an organization's overall information system by placing the one within the context of the other. GISs offer particularly exciting possibilities because they represent a real expansion and sense of growth in the area of management information.

In the next chapter, we will discuss spatial analysis and the cartographic capabilities—and limitations—of GIS in more detail.

FOUR

❖

Keeping the *G* in GIS:
Why Geography Matters

Geographic information systems represent a departure from typical analytical and policy development tools because of their explicitly geographic component. As the use of GIS becomes more common, the potential of this tool will become more readily apparent. One of the keys to unlocking the potential of GIS is a better understanding of geography among GIS users. This chapter attempts to shed light on the importance of geographic and cartographic principles to GIS implementation using examples drawn from public policy analysis and cartography. It recommends that organizations implementing a GIS develop a three-point approach to maximize the chances for success: (1) substantive expertise in the field of application; (2) knowledge of GIS techniques; and (3) understanding basic principles of geography and cartography.

Geographic information systems represent a significant departure from typical analytical and policy development tools because of their explicitly geographic component. As the use of GISs becomes more widespread, the potential of GISs in their various applications will become increasingly apparent. However, in their current state of development, the key to unlocking the full potential of GISs lies in a better understanding of geographic and cartographic principles among GIS users.

The purpose of this chapter is to demonstrate the importance of understanding basic geographic and cartographic principles in using a GIS as an analytical tool. Taking examples from the geographic literature, we will identify several problems that those unfamiliar with

GISs may encounter while using the technology, thus highlighting the importance of geography and cartography in the successful implementation of GISs. In particular, two themes will be covered: (1) the importance of scale and aggregation in spatial analysis and (2) the appropriate development and use of maps, especially with regard to generalization.

We will begin with a brief discussion of spatial analysis, and proceed to provide examples focusing on scale and aggregation and map use. We will conclude with a three-point approach to implementing GISs for analytical purposes, suggesting that the presence of substantive expertise, knowledge of GIS techniques, and the understanding of basic geographic and cartographic principles will improve the chances for an organization's acceptance and use of a GIS.

BACKGROUND

Although GISs have a history that dates back to the 1960s, we are a long way from unlocking the full potential of the GIS as an analytical tool. According to Goodchild and Getis (1991), while the potential of the GIS is vast, the current orientation of GIS applications is toward information and infrastructure management rather than spatial and policy analysis. In essence, as the authors have noted, the GIS is to spatial analysis what statistical packages (e.g., SPSS, SAS, etc.) are to traditional statistical analysis; thus it represents a powerful tool indeed.

However, most real-world applications of GISs rely primarily on the information storage and management functions of the technology. While there is nothing inherently wrong with using GISs in this way, the technology may be used to perform far more complex tasks in addition to the basic tasks of storage and retrieval. The key to using GISs to perform more complex tasks lies in an understanding of the basic geographic and cartographic principles that underlie spatial analysis, as well as data input and analytical output.

Geography is important to GISs for two major reasons. The first reason is that a sound understanding of basic geographic principles will increase the likelihood that GIS users will employ the technology in appropriate, logical, and ethical ways. The second reason is that the current emphasis on information storage and management undersells the value of the GIS as an analytical tool. By way of analogy, it has the effect of using a Cray supercomputer to perform simple arithmetic (processes that are well within its capabilities but not at all challenging), or owning a Maserati yet using it only to drive to the corner grocery store. A more prominent role for geography will unlock additional analytical capabilities, and thus make the technology even more useful.

We should not be surprised that the geographic potential of GISs is relatively underutilized. Many reports appearing in both scholarly journals and the mass media over the past several years have described the U.S. populace as largely geographically illiterate and have provided examples of the inability of students at all grade levels to locate familiar places on a map. But singling out map identification as an indicator of geographic literacy overlooks the breadth and depth of geography as a discipline and ignores geographic principles as a foundation for the development of the field. However, if we as a nation have difficulty with basic map identification, how can we expect to be knowledgeable about principles of geographic location?

Basically, geographic principles (and thus geography) help to explain and predict the location of people, places, and things. Consistent with this description of geography, Goodchild and Getis (1991) define spatial analysis as "a set of techniques whose results are dependent on the locations of the objects being analyzed" (1–2). That is to say, spatial analysis is appropriate in situations where results change when objects move. Goodchild and Getis give as an example the U.S. Center of Population, which has been moving westward over the years, following the migration of people from the Northeast toward the West and Southwest. When the United States was founded, its center was located near the eastern seaboard; today, the center of population is in Oklahoma.

Another example of spatial analysis can be found in the effect of the movement of the U.S. population on the location of congressional districts and the allocation of members of Congress among the states. For example, after the 1990 decennial census, the Sunbelt states of California and Texas added districts and representatives, while several midwestern states, such as Iowa, lost seats.

It is important to note, however, that the statistical analyses with which most of us are familiar are unaffected by movement—they are aspatial. For example, according to Goodchild and Getis, the average income of the U.S. population is unaffected by the states of residence of U.S. citizens.

Although the current orientation of GISs in practice is on information storage and retrieval, particularly information related to infrastructure, GISs are capable of far more complex tasks. Among these tasks, according to Goodchild and Getis, are map measurement, particularly area measurements, which are simple to conceptualize but difficult to execute manually. But GISs can also be used to perform even more challenging analyses because of their vast capabilities for data integration and manipulation. Such analyses include those needed in the development of global science, Goodchild and Getis have observed.

The point to bear in mind, however, is that in order to maximize our use of GISs, it is necessary to understand basic geographic and cartographic principles. The following sections suggest how a sound understanding of these principles can improve our analysis and outputs.

TOBLER'S FIRST LAW OF GEOGRAPHY

Tobler's (1970) first law of geography boldly asserts that "everything is related to everything else, but near things are more related than distant things" (236). This is the basic assumption that underpins geography as a discipline. Geographers devote their professional lives to looking for these relationships, which take the form of geographic or spatial patterns covering the face of the earth. For example, when we read newspaper accounts about migration to and from the "Sunbelt," or flooding in the "Midwest," we have a common understanding of what region the reporter is describing. Our understanding is based on our recognition of relative uniformity within the region, and some contrast between the regions.

Specifically, the word "Sunbelt" conjures up images of—what else?—warm, sunny climes; leisure activities; retirees; and in the 1970s and 1980s, job growth. The term "Midwest" brings to mind amber (and green) waves of grain and other crops, solid American values; the pioneer spirit; and in recent decades, job loss. However, in spite of the similarity of our generalized images of these regions, when we get down to specifics, differences of opinion arise and may become heated.

Take the case of circumscribing the Midwest. There is probably widespread agreement about including the states of Illinois, Indiana, Iowa, Michigan, Ohio, Wisconsin, Missouri, and Minnesota. Some people argue, however, that Kansas, Nebraska, and Oklahoma should also be included. Others suggest that South Dakota and Kentucky or Pennsylvania (sometimes in whole, sometimes in part) should be included as part of the Midwest. A sociologist friend recently defined the Midwest as stretching from Cleveland to Colorado. In the geographic literature, Zelinsky's (1980) Midwest includes all or parts of the states of Ohio, Michigan, Indiana, Illinois, Wisconsin, Missouri, Iowa, Missouri, Minnesota, South Dakota, Nebraska, Kansas, and Oklahoma. By contrast, DeBlij and Muller (1992) consider a large proportion of the states of Illinois, Indiana, Ohio, and Michigan to be more appropriately classified as "the Anglo-American Core," which by their definition is "synonymous with the American Manufacturing Belt" (215). Finally, Garreau (1981) includes most of Illinois in "the Breadbasket," but classifies much of Indiana and most of Ohio and Michigan as parts of "the Foundry." Southern Indiana and Illinois, by the way, are part of Garreau's "Dixie."

How can it be that a region that we talk about so casually can be so difficult to define? There are two major reasons, one relating to the difficulty of developing meaningful taxonomies or classifications, the second relating to Tobler's first rule of geography.

What we are seeing in the classifications of Zelinsky (1980), DeBlij and Muller (1992), and Garreau (1981), is their idiosyncratic definitions of what constitutes a region. Gould (1983) observed that

> taxonomy is often regarded as the dullest of subjects, fit only for mindless ordering and sometimes denigrated within sciences as mere "stamp collecting.". . . If systems of classification were neutral hat racks for hanging the facts of the world, this disdain might be justified. But classifications both reflect and direct our thinking. The way we order represents the way we think. (72)

It is likely that the three classifications of "midwestern" states described above reflect differences in the criteria for classification that the authors have chosen to define their regions. We are seeing differences in the way these authors think. Anyone who has driven throughout this region can testify to the existence of not only miles and miles of productive farmland but also factories—some of them still in operation, others long closed. Given individual differences, we should not expect an exact match in the development of a regional classification scheme.

However, whereas Gould's explanation gives us a general clue about how three classification schemes could be so different, Tobler's first law of geography provides more specific insight. To reiterate, Tobler has asserted that "everything is related to everything else, but near things are more related than distant things" (1970, 236). Tobler's law gets to the heart of an issue of crucial importance to geography, namely, scale.

Several examples will help to illustrate Tobler's law, which applies to both history and geography. Taking a historical example, it has been noted that a weather forecaster would have a very good probability of predicting tomorrow's weather if he or she did nothing more than say that today's weather pattern would continue through tomorrow. However, if that same forecaster predicted that today's weather pattern would prevail at a date six months from now, the likelihood of an accurate prediction would be minuscule, particularly in a region like the midwestern United States where temperatures can fall well below zero in the winter and soar to the 100 degree (Fahrenheit) range in the summer. And while predicting today's weather pattern for a date three months, one month, or one week from today might be closer to what we would actually see on those dates than the weather we might see on a date six months away, the probability of accuracy would not be nearly as great as the initial prediction that tomorrow's weather would be like today's.

Similarly, in the geographic mode, we can predict that today's weather in the town next door will be like the weather in our own hometown with a high probability of correctness. As we begin to predict our weather pattern as the prevailing pattern for another state, the likelihood of accuracy declines. For example, we might well expect that during a drive through Indiana to Illinois or Ohio we would experience little or no change in weather. On the other hand, if we were to travel to Maine, we might very likely see a difference in temperature, precipitation, and humidity. At the extreme, the difference between a typical hot and humid Indiana summer's day and the frigid Antarctic on the same day show the folly in trying to extrapolate over large geographic areas.

We need not confine our illustrations to the physical. For example, consider the cultural differences that we see among people of various places. While Hoosiers (natives of Indiana) may seem a lot like Buckeyes (natives of Ohio), they will appear to be less like Mainers, and even less like the French, the Japanese, or the Maoris (of New Zealand).

And yet, consistent with Tobler's law, close to Indiana we would see many similarities. And even at great distances, we could find relationships between places. For example, the hole in the ozone layer over the Antarctic is usually attributed to the activities of people in industrialized countries. Hoosiers and Buckeyes, distant as they are from Antarctica, affect it.

Embedded within this notion that everything is related but nearer things are more closely related than distant things is the hint that transitions occur across space. These transitions are rarely smooth and clearly defined. A visit to Maine demonstrates this fact. Maine is located adjacent to the Canadian province of Quebec, which has French as its official language. Most Quebecois (residents of Quebec) are descendants of French immigrants. In spite of the fact that there is very clear and definite line on the map separating Maine from Quebec, the cultural line on the ground is, as Goodchild (1988) might describe it, "fuzzy." A large proportion of the Mainers who live near the border of Quebec are self-described Francos—that is, Mainers whose ancestors were French. And although the concentration of Francos declines as one leaves the Quebec border and drives "Down East" to the Maine coast, the descendants of French immigrants can be found throughout Maine and recognizably into Massachusetts. By the time one gets to Indiana, people of French ancestry are few and far between, having been replaced by descendants of British, German, Eastern European, and African peoples.

Thus, drawing a political boundary is a relatively easy task, once the border countries agree on who gets what. Such a border will have exact geographical coordinates. On the other hand, trying to provide an accurate, understandable, meaningful map showing the attributes of the land and its people is much more difficult because of the problems created by transitions, such as the one described above.

PROBLEMS OF REPRESENTING THE REAL WORLD ON A MAP

The problems of representing ground truth (what objectively exists in reality) are well known to cartographers. Imhof (1963) described cartography as "a technical science with a strong artistic trend" and observed that

> theoretical cartography is not yet sufficiently characterized. The motif or theme of the map is the earth's surface. The essential element of any drawing (or pictorial design) is observation. The people who are drawing the surface of the earth must observe and study it. The geographer also does this; this is a geographical task. In this respect, there is a very close relationship between geography and cartography. To a certain degree the cartographer is a geographer; he is a graphic geographer or a geographic artist. As an applied science, theoretical cartography bridges the

connection between techniques to art, to all the different sciences that are concerned with the study of the phenomena of the earth's surface. (14)

Clearly, the cartographer's task is not an easy one. And in "the olden days," when a cartographer's tools emphasized pen and ink, most people found the cartographer's craft daunting. Few took the challenge to learn the intricacies of cartography. Nowadays, however, the introduction of GISs and computer-assisted design and computer-assisted mapping (CAD/CAM) enables anyone to make a map.

It would be a big mistake, however, to assume that the ability to make a map makes one a cartographer. Even in the early 1960s, Imhof raised concerns about the use—and potential misuse—of "technical aids" in the hands of people unfamiliar with the basics of cartography and geography. He noted that "the cartographic problems concerned with the graphic composition and design of the map have increased in complexity, in spite of the new technical aids" (1963, 15). Imhof identified five problem areas in modern cartography: (1) generalization, (2) coordination, (3) visual effect, (4) abstract and direct pictorial representation of terrain, and (5) thematic mapping.

Generalization

All maps are generalizations of the areas they represent. Imhof noted that even if we were to take an aerial photograph of an area, then reproduce it at the size we desire, we would find it difficult to identify all the relevant features for the task at hand. According to Imhof, it is, therefore, the cartographer's task to "transform, emphasize, eliminate, summarize, exaggerate and enlarge certain things" in the map in order to convey an image that is usable (17). This is the process of generalization. As Imhof went on to say, "The solution of a mathematical task is either right or wrong, but the solution of a cartographic task can be evaluated only in degrees of good or bad" (17). At one time or other, most of us have had experiences with maps that were inaccurate, incomplete, or ambiguous. If you were using such a map to get from one place to another, you might find key landmarks missing, roads where none exist, or other points of confusion that would make using such a map frustrating. Generalization must be done with great care.

Coordination

Once the cartographer has chosen the features that he or she wishes to represent and a way to represent them, it is necessary to coordinate the resulting elements. The idea is to create a map that conveys information in a usable form. This means that roads, rivers, and political boundaries must be distinguishable from each other. It means that topographic features, where present, must be recognizable. It means that the names of towns, cities, states, and other places must not obscure other important information and must be unambiguously related to the correct place. As Imhof (1963) noted, it is possible to design a map with excellent elements only to find that the neglect of the overall effect results in a poor map (17).

Visual Effect

Closely related to the problem of coordination is the problem of visual effect. What does the map look like once it is completed? Is it neat or cluttered? Are the symbols helpful or confusing? Does the map convey the message it was designed to convey? In assessing the visual effect of a map, it is crucial to keep in mind that a map is designed to convey information or ideas to others, and that it is not enough that the mapmaker understands his or her own creation. The visual effect must be clear to the audience for whom the map is intended.

Abstract and Direct Pictorial Representation of Terrain

The representation of terrain is an important aspect of every map. Imhof (1963) noted that such representation "forms the look of many maps more than anything else" (18). According to him, the representation of terrain on maps must fulfill two requirements: (1) it must represent surface shapes geometrically, and (2) it must present a readily recognizable pictorial representation of the terrain. Typically, contour lines are used to show variations in elevation. Unfortunately, such lines require careful inspection to reveal the lay of the land. In some cases, cartographers use shading and other salient visual devices to render a picture that more prominently depicts hills and valleys.

Imhof also noted that, whether the cartographer uses contour lines or shading, "any cartographic terrain representation are [sic] somewhat fictitious in nature" (1963, 14). In part, the fiction is related to the fuzzy boundary and generalization problems that make it necessary to draw a fixed line to separate one elevation from another, when the transition is likely to be more gradual. The other part of the fiction is the image (created by shading, hachures, or use of colors) that the map reader sees and visually understands. This image may or may not accurately represent the terrain. What you see may or may not be what you get.

Problems of Thematic Mapping

Finally, Imhof described problems of thematic mapping. Thematic maps are typically designed to show the existence of a specific object or attribute across a defined area on the surface of the earth. Imhof gave the example of population. In some maps of this type, the cartographer places circles, dots, or some similar symbol of various sizes to correspond to the geographic location of cities, towns, or other places. When such a map is well done, the map reader gets a good idea of the sizes of various places.

Another method frequently found in thematic maps is the use of shading to show the incidence of particular objects or attributes. The usual rule in such a practice is the following: the higher the density of the object or attribute, the darker the shading. A serious problem may arise in using the shading method. Because the areas into which the map is divided for shading purposes are typically administrative or political districts (states, counties, and so forth), the resulting image may be misleading. For example, if one were to take the cities in Illinois and shade them according to their level of poverty, the city of Chicago (and most cities) would be more or less in a middle range. Yet a drive through Chicago quickly reveals that there are vast differences in the wealth of city residents, from the millionaires on the Gold Coast of Lake Michigan to the poverty of the Cabrini Green housing project nearby. The typical shading found on many thematic maps causes the loss of much important detail. Imhof has contended that many such maps exhibit cases of "shocking ambiguity" (1963, 21).

The dilemma surrounding the use of a GIS is that the system allows any reasonably intelligent, computer-literate person to make a map, regardless of his or her knowledge of the difficulties associated with cartography. Even more seriously, an individual without cartographic training who sits down at a GIS workstation to make a map may very well be unaware of his or her limitations. The ease with which a person armed with a GIS can make a map may lull the individual into a false sense of security as far as expertise is concerned. It is important to realize, however, that limitations do exist. Just as no reasonably intelligent person who has mastered Microsoft Word on his or her personal computer would automatically assume that he or she has suddenly become capable of writing a novel with the staying power of A Tale of Two Cities, neither should the novice who has mastered the use of a GIS assume that he or she has suddenly become an expert cartographer.

In trying to describe the basic problems inherent in undertaking geographic analysis and cartographic tasks, we have provided some general examples of the difficulties that one might encounter. Below, we provide examples designed to illustrate the complexity of geographic analysis and cartographic design and use. The examples we present come from the public arena, but appropriate examples can be found in the private sector as well. In the public sector, the potentially widespread public impacts of analysis and resulting policy choices pointedly reinforce the importance of a sound understanding of geographic and cartographic principles within the framework of GIS implementation.

Practical Examples

The first example comes from Chicago and its surrounding counties, which are united by a jointly funded transportation system, the Regional Transportation Authority (RTA). The example recounts a debate among the members of the RTA about "regional equity," a policy theme centering on the geographic sources of revenue versus the geographic distribution of funding for transportation services within the RTA's service-delivery area (Obermeyer, 1990b).

The RTA was formed in 1974 by the Illinois legislature and won approval in a state referendum. The purpose of the RTA was to expand the tax base beyond the city of Chicago in order to support public transportation in the region, which covers Chicago and Cook County, along with the "collar counties": DuPage, Kane, Lake, McHenry, and Will.

Funding for the RTA comes from three sources. The first of these is fare-box receipts. The second is federal funds, such as those provided by transportation grant programs. The final source of funds comes from sales taxes collected throughout the region. A crucial point in analyzing regional equity in this example is that the sales-tax rate is not uniform throughout the region: Cook County residents pay a full percent in sales tax to support the RTA, while collar county dwellers pay only a quarter of a percent in sales tax.

Originally, the RTA directly operated three separate transportation services under the auspices of three different service providers: the Chicago Transit Authority (city buses and rapid transit), a commuter rail network, and a suburban bus system. A budget crisis in 1981 resulted in a reorganization of the RTA that formally separated the services and made the RTA an administrative, rather than an operational, body. After the reorganization, the Chicago Transit Authority continued to be known as the CTA. The commuter rail network became "Metra," and the suburban bus system became known as "Pace." The RTA became an administrative overseer, rather than a direct service provider.

There were two major sources of conflict in this case. The first source of conflict was the uneven density of transportation services in the service-delivery area. Transportation service is much more readily available in the city of Chicago and Cook County than it is in the collar counties. The second conflict arose because residents of the collar counties perceived themselves as "forced riders," paying for a service for which they had relatively little use, as evidenced by low ridership among residents of the collar counties.

The regional-equity debate began with the drafting of the RTA's budget for fiscal year 1986 as the transportation services jockeyed for position in order to maximize their shares of the sales-tax revenues collected by the RTA. Metra initially raised the issue, arguing that the suburbs provided 60% of the RTA's funds from sales taxes, while receiving less than 40% of this revenue to cover the operating expenses of the two services—Metra and Pace—providing public mass transportation in the suburbs. Predictably, Metra's suggested solution to this problem was greater funding for itself and Pace.

The CTA justified its relatively larger share of the sales-tax revenues by noting that 37% of the trips taken by suburbanites were via CTA vehicles, while only 21% of the sales taxes collected in the suburbs to help fund the RTA went to CTA operations. Pace responded by formally declining to become involved, suggesting that the regional equity theme was a smoke screen, that equity was in the eye of the beholder, and that Metra's primary objective in raising the issue was to maximize its share of RTA revenues.

Both Metra's and the CTA's analyses were based on verifiable facts and were therefore valid in that respect. The difference, however, is that each relied on a different set of review criteria focused at different levels. Metra's analysis was valid at the regional, aggregate scale. The CTA's analysis was essentially focused at the local level. Not surprisingly, each came up with a different solution.

We can see the difference that scale of analysis makes even more clearly by comparing Metra's regional, aggregate analysis with an analysis based at the individual, disaggregated level. To reiterate, Metra's analysis charged regional inequity based on the fact that the suburban areas of metropolitan Chicago contributed 60% of the RTA's operating funds through sales taxes collected, while the two suburban-based transportation services received less than 40% of the RTA's revenue from sales taxes to support operations. At the individual level, we have a starkly different situation: suburban residents paying only a quarter of a percent in sales taxes to the RTA, while city of Chicago residents are paying a full percent in sales taxes, or four times the amount required of suburbanites. In this instance, changing the scale of the analysis from the regional, aggregate level to the individual, disaggregate level changes the results of the analysis dramatically.

Of course, it is important to point out that the idea behind the creation of the RTA was to expand the public transportation tax base to include suburban areas. Based on its operating charter, it may be argued that it was never intended that each geographic area should get back in services exactly what it had paid in taxes: if that had been the intention, there would have been no need for an RTA. At the very least, the funding and spending mechanisms would have been designed differently.

If we carry this theme to its logical conclusion, we must also consider the effects of scale on the development and implementation of policy. An excellent example of this theme is found in Paul Knox's (1988) article, "Disappearing Targets? Poverty Areas in Central Cities." In his examination of poverty areas within U.S. cities, Knox documented the existence of variations in both the nature and the intensity of urban poverty, attributing these differences to the "differential imprint of economic and demographic change among cities and regions of the country" (502). He suggested that the policies established under the Reagan Administration and maintained throughout the Bush Administration treated poverty as an individual problem and delegated to the states the responsibility for providing

AFDC payments to the poor living within their borders. As a remedy, Knox recommended that public policy be targeted spatially and be "customized to address the problems inherent to specific, localized cycles of deprivation" (506). Such targeting might include locating well-baby and family medical clinics or job training and development centers in the neighborhood or providing special incentives to businesses to locate in these areas as a means of enhancing local economic opportunities.

The more general theme of Knox's article is that our definition of problems affects the way in which we address them. More specifically, Knox suggested that the scale at which we define problems determines the strategies we adopt to resolve them.

While GISs give us enhanced operating capacity by increasing the quantity of data that we can analyze, they cannot by themselves develop appropriate modes and scales of analysis in a specific circumstance. That is for the project analyst or manager to do. It is therefore necessary that individuals responsible for developing approaches to analyzing geographic information have a sound understanding of geographic principles and substantive expertise in their organization's mission in order to prepare them to perform this assignment competently.

MAPS AND THEIR APPROPRIATE USE

A second area of concern that we will examine is the appropriate use of maps and the concomitant need for cartographic expertise. This theme is worthy of attention because of the important role of maps both as inputs into and outputs from GISs. As inputs, existing maps may be scanned or digitized to provide a basic framework for other data in a system. As outputs, maps may be used by a variety of people for a number of different purposes, some of which may not have been intended by the creators of the map. More seriously, some of these unintended uses may be entirely inappropriate.

In his book *How to Lie with Maps*, Mark Monmonier (1991) provides a thorough discussion of the use and misuse of maps. Specifically, he notes that because of the availability of personal computers and other electronic mapmaking technologies, mapmaking has become available as a tool to those who have the hardware and software or who find themselves assigned to this task on the job. Not surprisingly, many of the people who now assume the role of cartographer, or find themselves thrust into that role, have no training whatsoever in cartography. Consequently, the probability of inappropriate map creation or use has been increased in recent years. Certainly, the proliferation of geographic information systems exacerbates this situation by increasing the ease and speed of map creation and use.

This is clearly a problem. Monmonier (1991) emphasizes its seriousness:

> *A single map is but one of an indefinitely large number of maps that might be produced for the same situation or from the same data.* The italics reflect an academic lifetime of browbeating undergraduates with this obvious but readily ignored warning. How easy it is to forget, and how revealing to recall, that map authors can experiment freely with features, measurements, area of coverage, and symbols and can pick the map that best presents their case or supports their unconscious bias. Map users must be aware that cartographic license is enormously broad. (2)

When using GISs to make new maps and when using existing maps to create other maps, caution is in order. Both inadvertent mistakes and intentional bias may make a specific map unusable for a variety of purposes. However, even mistake-free and unbiased maps may be used inappropriately.

Gersmehl (1985) provides an excellent example of this situation in which he describes how one of his own maps was misused. In 1977, Gersmehl published a set of coarse-resolution dot maps showing the general distribution of soil orders in the 48 contiguous states of the United States. His map showed the presence of a general class of soils, histosols, in several western states. These histosols represented some widely dispersed peatlands as well as "muck." Although they both qualify as histosols, peat and muck are different: peat is valuable as an energy resource; muck is not. Herein lay the basis for misinterpretation and misuse of Gersmehl's map.

Attempting to identify potential energy resources in the United States, the U.S. Department of Energy (DOE) came across Gersmehl's map and included the information provided on it in the agency's map of commercial energy fuel. In compiling this map, the DOE failed to recognize that the class "histosol" was a general soil type that included everything from peat to muck. Instead, the agency classed all the histosols as peat, an assumption that resulted in a map showing a much greater amount of this valuable energy resource than actually exists.

The result was a map that was factually incorrect through no deliberate intention to falsify either on the part of the creator of the original map or the developer of the derivative map. Gersmehl related his surprise and concern about the misuse of his map and concluded that "a person who puts information on a map has a duty to be fair to the data, to be clear to the map reader, and to try to anticipate the ways in which a third person may be affected by a foreseeable interpretation of the map" (334).

This caveat seems to put all the responsibility on the cartographer not only to make a map that is accurate and readily understandable but also to try to anticipate the many ways in which a specific map might be misused. Unfortunately, Gersmehl's emphasis on the cartographer seems to let the map user off the hook. Not only must the mapmaker be aware of the potential misuse of maps but also the map user must bear responsibility for using existing maps appropriately. This requires some basic understanding of cartography along with a willingness to make the effort to consult with the original cartographer regarding appropriate use of a specific map if there is even the slightest question about the information transmitted via the map.

If caution is appropriate in the use of paper maps, then caution is most certainly required in the use of GISs. The ease of use of the technology, coupled with its speed, means that GISs can encourage the proliferation of cartographic errors at a rapid pace in the hands of inexperienced or unknowledgeable users.

A THREE-POINT APPROACH TO RESPONSIBLE GIS APPLICATION

The importance of the GIS as a tool in policy analysis is growing; and with improvements in the system's speed and capabilities and a reduction in the price of this technology, its importance is likely to continue growing. The preceding discussion (of how changing the scale and aggregation of an analysis affects the results) highlights the crucial importance of an understanding of geographical principles before undertaking such an analysis. Similarly, the potential for misuse of maps suggests that a GIS user must understand the basic principles of cartography. In short, GISs may be used or misused. An appropriate balance of skills and knowledge is the only hedge against misuse and is the key to unlocking the full potential of GIS technology.

In order to unlock the full potential of GISs and to use the technology in an appropriate manner, we recommend a three-point approach to effectively applying GISs. In order to maximize the functions of the technology, GIS users must possess the following three attributes: (1) substantive expertise in the field where the GIS is to be used; (2) knowledge of GIS techniques; and (3) an understanding of geographic and cartographic principles, at least in rudimentary form.

Each application of a GIS is unique. To begin with, each application occurs within the framework of a specific organization mission and has specific geographic boundaries. Therefore, for example, using a GIS to identify potential toxic waste dump sites throughout the United States is very different from using a GIS in Cincinnati for infrastructure management. Each application requires specific substantive expertise related to the mission of an organization and the specific task at hand.

Presumably, the substantive expertise already exists within the organization adopting a GIS. It is crucial to make use of that expertise, to find a means to keep the organization's substantive experts in the GIS loop to help develop data specifications and sources, then to monitor the implementation, and finally to evaluate the results. These experts are in the best position to determine the validity of results and recommend necessary changes.

Knowledge of GIS techniques, and specific knowledge of how to operate the system chosen is absolutely necessary. The individual in the GIS driver's seat must know not only the commands to operate the system, but also the capabilities and limitations of the system. It is this individual's job to assure that all tasks undertaken using the GIS are within the capabilities of the system.

Finally, the importance of understanding basic geographic and cartographic principles cannot be overemphasized. Critics have argued that this third point is exclusionary. That is to say, it sets up a standard that allows only geographers to handle an organization's GIS. It is not our purpose to recommend a kind of technical "litmus test" for operating GIS technology. We argue, instead, that whomever an organization charges with the responsibility for operating its GIS must have sufficient geographic and cartographic background so that problems of the type described in this chapter will not interfere with the successful application of GIS technologies.

But how can an organization making use of a GIS assure that it meets all three of these requirements? Modern education, with its emphasis on specialization virtually assures that no single individual will possess substantive expertise in the organization's function, knowledge of GIS techniques, and an understanding of geographic and cartographic principles. More likely is a situation wherein these three areas may be covered by three (or perhaps even more) different people.

Obviously, this situation requires extensive and careful coordination to make certain that all relevant professionals are kept in the loop. Such coordination begins with the realization and acknowledgment that all three functional areas are of equal and crucial importance. Without the active involvement of each area and mutual respect between and among the experts, there is a danger of an unsuccessful implementation. Dominance by each of the areas brings its own particular difficulties.

For example, if the substantive experts fail to acknowledge the expertise of the GIS professional, they may insist on operations that are inappropriate to the system and database that they have available. If they do not respect the knowledge of those with expertise in geography and cartography, they may make mistakes related to scale or cartographic features and obtain results such as those described by Gersmehl.

Similarly, if the GIS expert fails to respect the substantive expertise and geographic and cartographic knowledge of his or her colleagues, he or she may fail to listen carefully to the information and recommendations of colleagues and may consequently fail to incorporate such knowledge into the system. This can result in a system that is poorly or incorrectly specified and that consequently will not perform up to its potential. Anyone who used computers in the days of punch cards and "consultants" has experienced this potential problem first-hand.

Finally, if the person who is knowledgeable about geography and cartography does not respect the expertise of the substantive specialist and the GIS professional, he or she may miss crucial details that are unique to the specific application involved. Again, the project will suffer.

Each organization will have to develop its own strategy for coordinating these three groups. Regular meetings and discussions may help promote communication and cooperation. In addition, providing training and education via workshops or enrollment in relevant courses in nearby colleges and universities may also help the individuals involved gain a basic understanding of all three elements of a project.

CONCLUSIONS

Effectively applying a GIS technology is not an easy task. It requires an understanding of the substantive application area, a thorough knowledge of the GIS employed, and an understanding of the basic principles in geography and cartography. Of these three components, the third is least likely to be recognized as necessary. And yet, as we have seen, it is crucial to the development and successful implementation of a GIS used for analytical purposes.

When organization managers think about implementing a GIS, their first thought is to acquire a system and hire a technician. With the current emphasis of GIS applications primarily on data management, in the short run this is perhaps not a very serious problem. However, it does mean that the GIS may not be used as fully as possible, and, therefore, that the organization may not accrue the full benefits of owning the technology.

In order to achieve the full benefits of owning a GIS, it is necessary to have a substantive expertise, GIS skills, and an understanding of basic geographic and cartographic principles. While we can only suggest how this triumvirate of knowledge may be achieved, each organization must find its own way. The first step, however, is recognizing the need.

❖

FIVE

Bureaucratic Factors in the Adoption of GISs

Gaining the acceptance of or commitment to new technologies and innovations from members of organizations can often be a frustrating, time-consuming process. The achievement of this goal depends on a number of factors, among them technical, financial, and political conditions, as well as bureaucratic structures that influence the adoption of new innovations, generally, and GISs, specifically. In this chapter, we will argue that because of an organization's reliance on "standard operating procedures," on the one hand, and professional training and socialization, on the other, the adoption of GIS technologies is often hampered. In general, organization members tend to favor the status quo, while administrators often screen out the geographical components of professional tasks and rely on nongeographic information systems. This argument is based on a theoretical understanding of bureaucracy, organizational decision making, and the search for information used by organizations in the decision-making process. We will offer support for this argument from a study of the use of a basic form of geographic information—maps—by public administrators and planners.

Make no mistake, the potential adoption of a new technology, such as a GIS, represents a serious threat to the status quo of many organizations. The wide range of applications that a GIS can perform, coupled with the tremendous effect on altering the nature of work for many employees, has the potential to upset well-defined task relationships and lines of responsibility. As a result, the adoption of a new technology is often subject to some of the most severe resistance on the part of organization members. People feel threatened by a new technology and, as a result, are often inclined to resist its introduction. Because of the general reluctance on the part of organization members to accept technological change, much has been written on this topic in an effort to smooth the difficult transition period from the introduction of a new technology to its institutional and general use.

In this chapter we will focus on the decision by an organization to adopt a GIS, and we will take the position that the bureaucratic nature of many public and private agencies influences adoption of any innovation. Adoption of GISs by public agencies poses special problems, since, as an innovation, the GIS is both new and, to public administrators, uniquely different from their traditional professional tools. This argument is based on a theoretical understanding of bureaucracy, of decision making by organizations, and of the search for information used by organizations to make decisions. Specifically, this chapter asks the question, "How do bureaucratic factors influence the adoption of an innovation by private and public organizations in the United States?"

Our primary hypothesis is that bureaucratic factors play a more or less predictable role in the adoption of GISs by public organizations. Specifically, we suggest that as the result of use of "standard operating procedures" by organizations on the one hand, and professional training and socialization on the other, public organizations favor the status quo, while public administrators screen out the unfamiliar (in this case, geographic) components of professional tasks and problems and rely on their traditional tools (words and numbers), thus overlooking spatial patterns as a possible basis for analysis and communication. Therefore, GISs face two distinct impediments to application in the public sector: organizational bias for existing tools on the one hand, and professional bias against geographical analysis and communication on the other.

This chapter is organized as follows. It begins with a discussion of geography and the public sector. The next section focuses on pertinent organization theory and contains two subsections. The first subsection reviews organizational decision-making. The second subsection covers an organization's search for informational input into decisions, focusing specifically on the effects of (a) an organization's use of standard operating procedures and (b) professionalism. Through a preliminary study of recent volumes of professional journals of public administration and planning, the following section provides evidence to support our hypothesis regarding organization bias. The final section offers conclusions.

The decision to adopt a GIS involves a process whereby an organization (1) identifies a need, (2) searches for alternative methods or tools to satisfy its need, (3) chooses one of the alternatives it has found, and (4) implements its choice. This model emphasizes the search for alternatives and the choice among identified alternatives, recognizing that the GIS is but one of several types of automated information systems available to serve the public need for information and that the GIS competes in the marketplace against nongeographic information systems (e.g., MISs).

An exploration of the factors in bureaucracies and organizations affecting the adoption of GISs by public agencies gains significance as the thrust toward enhancing applications of GISs in the public sector builds steam. Already, geographers and GIS specialists have raised initial concerns regarding the role that geographers can play in promoting GISs (Dobson, 1983b; Kellerman, 1983). Dobson argued that geographers "should assume leadership to assure that the new spatial analysis techniques are used correctly and efficiently" (349), while McNally (1987) viewed the problem of promoting GIS as a problem of marketing and said, "We [geographers] just have to learn to adapt the product to fit the needs of the marketplace" (391). Muehrcke (1981) suggested that "greater attention . . . be directed at discovering ways of raising the cognitive level of all map users" as a means to promote GISs (403).

Most geographers and GIS specialists have emphasized technical, financial, and political impediments to the adoption of GISs (Cromley, 1983); others, however, have referred to the role of organizational behavior in the adoption of GISs. For example, without elaboration Muehrcke and Dobson (1983a) noted the presence of systematic professional bias in the acquisition and use of information. As a potential obstacle to implementation of GISs, both Mead (1981) and Chrisman (1987b) mentioned organizations' preference for the status quo; and Chrisman specifically related this preference to professional bias (35). Similarly, Portner and Niemann (1983) asserted that variations in professional values regarding land information have had a negative effect on implementation of GISs in Wisconsin. Organization theorists (Cyert & March, 1963; Downs, 1967b; Simon, 1945/1976) have explored more thoroughly the link between professional bias and the acquisition and use of information for decision making by organizations. This research explicitly links organization theory and GIS case studies in order to shed light on the adoption of GISs by public agencies.

So long as promoters of GISs limit specification of the GIS adoption issue to technical, financial, and political factors, they will seek to expand applications exclusively through modifications of these factors. If, as we suggest, influences within an organization play a significant role in the adoption of GISs, then no amount of improvement in financial and technical incentives or in the political environment will facilitate and expand the application of this important technology to its full potential as a public policy tool by public organizations.

(The problem of GIS adoption is significant in both the public and private sectors. It will become evident that many characteristics of organizations transcend public and private dimensions. Therefore, lessons learned from an exploration of bureaucratic factors in the adoption of GISs may be relevant to the expansion of GIS applications in both the public and private arenas.)

THEORETICAL FRAMEWORK FOR ANALYSIS OF GISs IN ORGANIZATIONS

Theoretical research on organizations supports the argument that there are attributes intrinsic to the bureaucracies of most organizations that influence the adoption and use of innovations in general, and that may therefore apply to GISs in particular. Key among these attributes are: (a) the use of standard operating procedures in the organization, which tends to predispose the organization to continue past practices while limiting the organization's search for information to only the familiar, and (b) the role of professionalism and professional training, which tends to limit systematically the information search and the professional tools selected.

Many organization theorists begin with the assumption that organizations generally function in a manner consistent with bureaucratic principles. The term "bureaucracy" has developed a negative connotation in current usage. Indeed, the word is often paired with such pejorative adjectives as "slow," "cumbersome," "inept," and "corrupt" (Goodsell 1983/1985). In its more technical usage, however, "bureaucracy" is the word that is used to identify a particular type of structure in an organization.

The ideal bureaucracy, as described by Max Weber (1946), may be either public or private and may be developed as a means of rationalizing the performance of the organization and, thereby, of improving its efficiency of operation (rationalizing in this sense includes the notions of routinization of the organization and standardization of operating procedures). Weber asserted that the "fully developed bureaucratic machine compares with other organizations exactly as does the machine with nonmechanical modes of production" (214). In theory, the ideal bureaucracy is characterized by precision, speed, absence of ambiguity, knowledge of files, continuity, discretion, unity, strict subordination, and reduction of friction as well as of material and personal costs. In reality, the gains in efficiency made possible by a bureaucratic arrangement have costs, namely, inflexibility created by the organization's bureaucratic structure.

ACQUIRING AND USING INFORMATION IN ORGANIZATION DECISIONS
Bounded Rationality

The starting point for discussions of decision making in organizations is the notion that the decision-making process is rational (Danziger, Dutton, Kling, & Kraemer, 1982). Implied in this notion of rationality is the requirement of complete and accurate information: the truth, the whole truth, and nothing but the truth. Organization theorists have come to realize, in recent years, that the information requirement so necessary to truly rational decision making is rarely obtained in practice. Consequently, discussion by these theorists has shifted to a phenomenon that Simon (1945/1976) dubbed "bounded rationality."

Bounded rationality suggests that the capacity of decision makers to be fully rational is bounded by three factors: values, skills, and knowledge—all of which are, in part, an outgrowth of professionalism and professional training. "Values" of the decision makers implies, in part, commitment to attaining the best possible alternative. Not surprisingly, not all decision makers are committed to the same level of information search. Because decision makers have differing types and levels of "skills," they can influence the individual's ability to discover and recognize decision alternatives. Together, these factors limit the search for—as well as the acquisition and use of—information in the organization, with important consequences for change in organizations, and, by extension, the adoption of innovations. Current debate on the bounds of rationality centers on the concept of "search" (Downs, 1967b; March, 1986; Thompson, 1967, 1969), which is concerned with how bounded rationality limits an organization's search by systematically screening information used as input in problem solving and decision making.

According to Simon (1945/1976) and Thompson (1976), within the confines of bounded rationality, the best anyone can hope for is to "satisfice," that is, to set up a standard of "good enough" and accept the first alternative that meets it. It is commonly held that familiar alternatives win out under these circumstances (Cyert & March, 1963; Douglas, 1986; Steinbruner, 1974). This concept is profound because it explicitly states that decision makers rarely "optimize," that is, arrive at the best possible decision alternative. Rather, they seek satisfactory ("satisficing") alternatives that address the problem at hand to an adequate, if not complete, degree.

Standard Operating Procedures

As noted, the bureaucratic structure arose as a means of rationalizing an organization's performance and, thereby, of improving efficiency; this structure continues to be a dominant form in organizations. In most instances, organizations adopt standard operating procedures (SOPs) as one method to rationalize performance and improve efficiency. As an unintended by-product, SOPs also tend to screen information and, thus, to maintain the status quo.

The use of SOPs, while helping the organization to run smoothly, is fraught with difficulties. Standard operating procedures, according to Cyert and March (1965), "are the memory of an organization" (100–101). Rather than investing heavily in long-range planning, organizations tend to rely heavily on past practices, including both internal SOPs and general industry practices (which may be viewed as professional SOPs) to make decisions.

Standard operating procedures are based on an organization's records or files. According to Cyert and March, "The kinds of records kept tell a good deal about the firm's perception of its own internal structures and the kind of world within which it exists" (106). Moreover, an organization's records tend to limit the elements of the environment that the organization considers and uses to make decisions. As a result, an organization is inclined to respond in a preprogrammed way, resulting in actions better adapted to old problems than to new ones (Cyert & March, 1963; Gordon, 1986).

Internal SOPs are transmitted via instructions to new employees during on-the-job training, resulting in a strong bias to do things as they have always been done within an organization. Professional SOPs are transmitted primarily via specialized training and education, along with professional socialization and literature, and tend to encourage development of a particular professional orientation. The use of SOPs may be viewed overall as a biased screen in the acquisition and use of information by an organization. As James Thompson (1969) noted, "The bureaucratic routines, standard operating procedures, and the like are strong impediments to change" (54).

Training and Professionalism

Rules within an organization's SOPs may be introduced or reinforced by a means other than training within the organization—namely, by professionalism and by recruitment and selection of personnel. An important source of the bureaucracy's power is the professionalism of bureaucrats and their ability to guard their knowledge from outsiders. Each profession has a body of knowledge that is uniquely its own. In order to become a member of a specific profession, the hopeful candidate must successfully complete a well-defined rite of passage that ordinarily culminates in some sort of professional certification.

Professional organizations that fail to establish a certification process that is generally accepted by practitioners of the profession, and especially by those who hire its specific professionals, leave open the door for outsiders to encroach upon their professional turf. The planning profession is a good example of this phenomenon. In spite of the existence of a certification process for planners (administered by the American Institute of Certified Planners [AICP]), an individual need not be certified to seek, win, or hold most jobs in planning. Moreover, many planning positions do not even require an education or degree in planning. Thus, the AICP has had only limited success as a professional organization in carving a niche for professional planners in the job market. The experience of the AICP and the planning profession does not minimize, however, the important role played by professionalism and professional training. Indeed, the planning profession continues to work to achieve its objective of securing its niche.

Cyert and March (1963) emphasized the importance of professional training: "The 'craft' of new employees specifies how a job is to be performed. . . . When a business firm hires an accountant, a dietician, a doctor, or a sanitary engineer, it hires not only an individual but also a large number of standard operating procedures that have been trained into the new member of the organization by outside agencies" (105). Consequently, the kind of training that a bureaucrat brings to the organization is important.

In the ideal bureaucracy, individual bureaucrats are trained to perform their duties as defined within an organization. This specialized training imparts essential skills and knowledge to the individual bureaucrat, but also helps to shape his or her values. This has significant consequences for an organization's search, introducing a systematic pattern of bias in screening information based on professional training, socialization, and practice.

Theorists have applied a variety of names to this phenomenon. Merton (1952) credited Veblen with the term "trained incapacity," Dewey with "occupational psychosis," and Warnotte with "professional deformation." (364). The concept described by these terms suggests that an individual's training produces abilities that function as inadequacies, information screens, or blind spots. Quoting Burke, Merton noted that "a way of seeing is a way of not seeing—a focus upon object A involves a neglect of object B" (364). As Simon (1952) put it, "To the fireman, fires are the most serious human problem; to the health officer, disease, and so forth" (191). By extension, it can be argued that the administrator views administration (or management) as the most serious problem; the geographer, spatial patterns and analysis.

As Cyert and March (1963) noted, the person who gathers the information is also the first to condense it, evaluate it, synthesize it—and, finally, to transmit it. Information comes into an organization from many sources through a variety of transmission points. Each of these transmission points represents an individual within the organization who may condense and evaluate and perhaps alter information (either intentionally or inadvertently) before passing it along. As a result, it makes a difference not only who gathers information but also what his or her professional training is.

We may infer from the earlier discussion of training and professional socialization within the context of SOPs that specialized training influences the way in which an individual bureaucrat transmits information. For example, someone trained as a public administration generalist is likely to gather and process information differently from someone trained as a geographer. Specifically, the public administration generalist is unlikely to perceive and transmit the geographical dimensions of information. By extension, he or she may be unaware of the value and complexity of spatial analysis and communication, and therefore may not find the GIS an indispensable tool in the professional toolbox. Moreover, when purchasing the GIS, he or she may not have the training to use the technology appropriately or to its fullest capacity, as noted in the preceding chapter.

In addition to specialized professional training, organizations must also contend with the external pressure that "good business practices" exert on performance in the organization. These practices have a number of sources, including management consultants, trade and professional associations, and management literature. The combination of internal SOPs, professional training, and "good business practices" work together to form and reinforce the bounds of rationality: the values, skills, and knowledge of the bureaucrat. To the extent that bureaucrats in a particular setting share values, skills and knowledge, their decisions will reflect a systematic bias imposed by the shared bounds of rationality growing from professionalism.

MODELS OF AN ORGANIZATION'S SEARCH

The organization's "search" is a critical element in the acquisition of information and raises questions about how the search proceeds, what is (and is not) the object of the search, who searches and so on. There are many models of search by an organization. March (1986) contends that "Decision makers operate in a surveillance mode more than they do in a problem-solving mode. In contrast to a theory of information that assumes that information is gathered to resolve a choice among alternatives, decision makers scan their environments for surprises and solutions" (245).

Agreeing generally with the "scanning" model of search, Douglas's (1986) cognitive approach argues that decision makers take a taxonomic view of the environment, as a means of classifying relevant information by focusing on key players and events, to the probable exclusion of all else. In other words, the world is seen as fitting into certain preconceived patterns, known as taxonomies. This is consistent with Gould's assertion that classifications both "reflect and direct our thinking" (1983, 72). This approach is enticing because it allows the decision maker or information gatherer to make comfortable and familiar attributions about the environment (whether strictly true or not). For example, one alluring taxonomic view for a social conservative would be to examine the recent Los Angeles riots as evidence of the futility of past government policies of "throwing money" at social problems. Similarly, Steinbruner (1974) suggested that decision makers employ pattern recognition rather than exhaustive and deliberate searches designed to uncover the optimal solution to the problem. Linbloom's concept of "muddling through" offers an incremental approach wherein decisions or actions occur as a result of limited comparison of alternatives that build on past activity, with only rare full-scale changes in direction or approach (Cayer & Weschler, 1988).

According to this reasoning, organizations tend to exhibit bias in favor of continuing existing patterns of information search (a specific SOP). By implication, the geographic aspects of administration practice are likely to be neglected, even when the addition of geographic information would improve analyses and, by extension, decision making and problem solving. This bias toward past practice is a hurdle that GISs will face as momentum to expand applications into virgin territory grows.

THE SEARCH FOR INFORMATION AND A GIS

An organization's search for information has implications for the adoption of a GIS. As noted previously, a GIS is one of several information systems available to public organizations. When an organization determines that it has a need for an information system, it may choose a GIS or it may choose a nongeographic information system (Williams, 1987). In some instances, an organization may identify an explicit need for a GIS; however, it should not be automatically assumed that the only choice is a GIS or nothing. From this perspective, the choice of a GIS is more complicated, involving a perceived need (or lack thereof) by a decision maker for a geographic capability in an organization's information system.

Kraemer and King (1976) agreed that the kind of ISs available to organizations is important. Because ISs vary in their capabilities, they also vary in their specific utility to particular organizations. Perry and Kraemer (1979) specifically cited the importance of the nature of the innovation (such as a GIS) to its rate of adoption.

The decision to adopt an IS is not an easy one, and there is growing consensus that the organization itself is an integral component in the decision and its outcome. At one time, it was generally assumed that technical decisions, such as the adoption of an IS, were unbiased, rational choices; this assumption has since been shown to be incorrect (Danziger et al., 1982; Klosterman, 1987).

Current discussions on the adoption of GISs and other ISs include an explicit acknowledgment of the role of an organization's structure, procedures, and staffing patterns in technical decisions. Liley (1987) and Saarinen (1987) focused on an organization's structure as an issue, and Wunderlich (1986) specifically identified institutional obstacles, linking them with staff capabilities. Klosterman (1987) acknowledged the bureaucratic component in the decision to adopt an IS, while Kirby (1986) identified professional values as a factor in such decisions. Epstein and Duchesneau (1984) noted that public officials set their own information standards, which in turn influence their choice of a system. Other participants in the decision making process of an organization—such as elected officials, board members, and vendors—have also been cited for their roles in the final decision (Klosterman, 1987; Perry & Kraemer, 1979).

CASE STUDY: THE USE AND VALUE OF GEOGRAPHIC INFORMATION TO PUBLIC ADMINISTRATORS AND PLANNERS

Measuring the Value of Geographic Information

As we suggested in Chapter Three, an inherent difficulty in understanding the value of geographic information to a particular person or organization is finding an objective means by which to measure such value. We cannot observe directly the value of geographic information to an organization as we can observe the number of women the organization employs, for example, or the number of hamburgers it sells. Therefore, it is necessary to identify an acceptable surrogate by which to measure the value of geographic information to a specific group.

Because successful introduction of ISs has so often been predicated on user satisfaction with the technology, one means to measure the value of geographic information may be to demonstrate the use of geographic information, on the assumption that an organization that values geographic information will be more likely to use it. Perhaps the most common way in which geographic information is employed is in the production and use of maps. Therefore, the use of maps by an organization may serve as a valid surrogate means by which to measure the value placed on geographic information. That is to say, the higher the value placed on geographic information, the more prevalent will be the use of maps.

The Study

A review of professional journals for the public administrator provides initial evidence supporting the notion that professional training and socialization set in motion systematic professional biases toward specific types of information and against others as a means of analyzing and communicating information. As a preliminary test of this hypothesis, we examined pertinent literature in the fields of public administration and (for comparative purposes) planning to assess the use of maps by public servants with specific types of training. We reviewed two journals as representative of the professional orientations of public administrators and planners toward maps: *Public Administration Review* (PAR), a bimonthly publication of the American Society for Public Administration (ASPA), and *Journal of the American Planning Association* (JAPA), a publication of the American Institute of Certified Planners (AICP) and the American Planning Association (APA).

The review of the two pertinent journals *PAR* and *JAPA* covers the years 1985 through 1989, a five-year period. Our purpose in performing this review was to assess the degree to which practitioners in public administration and planning use maps, and, by extension, value geographic information. If the review were to indicate regular use of maps by public administrators, then it is logical to expect that a GIS would find ready application among public sector professionals.

If, on the other hand, the review were to indicate that regular use of maps by public administrators is rare, then it is reasonable to expect that the value of geographic information is low. By extension, application of GISs in the public sector may face intransigence from professionals or organizations at the implementation phase. Generally speaking, are public servants trained and socialized to appreciate the value (and therefore the use) of maps? The findings from this review suggest that they are not.

As we shall see, the two journals differ dramatically in their geographic representations. Planners and public administrators are trained within separate disciplines, and, not surprisingly, the APA's publication contains far more maps than the ASPA's journal does.

Moreover, both journals include information on authors, providing an opportunity to compare the use of maps by planners and public administrators.

For the purpose of this review, we have defined the concept of a map rather narrowly, excluding site plans and graphics such as aerial photographs that produce a map-like image. Obviously, our review suffers from some degree of subjectivity, reflecting our own perception of what does and does not constitute a map. Similarly, in order to assign only one discipline to each article, we have established some rather arbitrary decision rules for categorizing the profession of authors. For articles with two authors, the discipline of the first author carried; where there were three or more authors, the majority identification determined the assignment of the article to an appropriate disciplinary category. If a single author identified either planning or public administration as his or her discipline, along with another discipline, we classified the author according to the planning/public administration identification. Please note that this review included only major articles; it excluded book reviews, organizational news, advertisements, and so on.

We should also note that the two journals have somewhat different policies with respect to the inclusion of maps in articles. The instructions to authors of *JAPA* specifically encourage the use of maps and other graphics in articles for publication. *PAR* does not mention maps or other graphics in its instructions to authors, but its articles frequently include figures and tables. This suggests that while *PAR* does not encourage the use of maps and other graphics, neither does it discourage it. In fact, during the period covered by this study, maps have appeared in *PAR*. These differing policies are consistent with the hypothesis of this article, suggesting a greater awareness of geographic information and analysis among planning professionals (and others trained in the spatial sciences) compared with public administration generalists.

During the study period, *JAPA* contained 131 major articles, 50 of them (38%) with maps; *PAR* contained 248 articles, only two of them (less than 0.4%) with maps. During the 5-year study period, authors of six articles in *PAR* identified themselves as planners; none of their *PAR* articles included a map. Two authors of articles in *JAPA* identified themselves as public administrators; one of the two articles included a map. We see a similar phenomenon when we review the inclusion of maps by geographers in these two journals. Four articles in *JAPA* listed geographers as authors; three of them contained maps.

The single article authored by a geographer in *PAR* contained no maps.

For the 5-year period reviewed, one must go all the way back to January 1985 (the first issue of the period in question) to find a map in *PAR*. This issue was devoted to "Emergency Management" and presented more than 20 articles on that subject. Emergency management has been a subject of interest to geographers; and, in fact, one of the articles published in this special issue of *PAR* was written by a pair of geographers, Kasperson and Pijawka. Ironically, their article (Kasperson & Pijawka, 1985) did not include a single map.

In November 1986, *PAR* was devoted once again to a single topic, "Public Management Information Systems." None of the articles contained in this special issue specifically discussed GISs; and as noted, there were no maps.

The topics addressed in *PAR* often lent themselves to mapping. Several articles appearing in *PAR* could have been presented in a geographic manner to improve the transmission and comprehension of information. For example, articles comparing the states within the context of a specific public policy issue were common. This state-by-state information could have been readily transmitted to the reader via a simple choropleth (thematic) map to give the reader a visual geographic overview of the article's research findings. As the old saying goes, "A picture is worth a thousand words." This brief analysis lends credence to Cowen's (1987) observation that "maps . . . are of little value to decision-makers" in the public sector (339).

While this is admittedly a simplistic measure of professional assessment of the value of geographic analysis and communication, it is suggestive nonetheless. And while mapping is only the tip of GIS capability and usefulness, it is perhaps the most obvious application of the technology.

We must be careful, however, not to assume that all planners have the geographic awareness suggested by this quick review of *JAPA*. As previously noted, individuals working as planners may or may not have professional training or accreditation in planning. We may therefore expect some variation in geographical awareness even among individuals working as planners. And variation in geographical awareness among planners may have implications beyond merely the use or nonuse of geographic information. Without adequate training and professional socialization, misuse of geographic information becomes a distinct possibility.

While case-study evidence strongly suggests the importance of organization dynamics or bureaucracy in the adoption of ISs generally and GISs particularly; literature on GISs lacks broad-based studies specifically examining this link. There is a need for research to fill this gap.

CONCLUSIONS

Current directions in organization theory support the hypothesis that an organization's SOPs hinder adoption of any innovation. Moreover, theories of the ways in which organizations search for information suggest that organizations screen information based, at least in part, on their professional training. While acknowledging the role of organizations in the adoption of GISs based on numerous case studies, there is no evidence in the GIS literature of any systematic examination of the phenomenon. Yet, it is within the organization setting that GISs will compete against other types of ISs for adoption by public organizations. And the potential for difficulty does not stop with adoption of GISs.

Once an organization has adopted a GIS and proceeds to implementation, additional difficulties are certain to arise. Key among these is the potential for misuse of geographic information, which becomes increasingly likely as GISs become more readily accessible to users without specific training in the use of spatial information.

The solution is not to keep GISs from the more generally trained user, whose work may be enhanced with the addition of geographic information and analysis. A more hopeful, though admittedly more difficult, alternative is to develop GISs that reflect an understanding of the particular constraints imposed by bureaucratic behavior. At the same time, it will be necessary to provide adequate education and training to generalist-adopters of GISs. This education should specifically address the use and value of geographic information within an organization.

The evidence provided in this chapter suggests that there is a need for a systematic examination of an organization's setting as a means of increasing our theoretical understanding of the role of bureaucracy in the adoption of GISs. Such research will enhance the long-term effectiveness of GIS technology transfer systems through smoothing the process of an organization's adoption and use of a GIS.

❖

Economic Justification for GIS Implementation

Once an organization has begun to explore the implementation of a GIS, the next step is to examine the propriety of choosing this option. Typically, the benefit–cost analysis has been the primary technique for justifying a GIS, in both public and private sector alike. While the benefit–cost analysis is valuable, it must be performed with care, particularly in the public sector. The major reason for caution in using the benefit–cost analysis is that the public sector operates under economic constraints different from those in the private sector because of the absence of traditional pricing mechanisms. The public sector is not motivated by profit but by its mission to serve the public interest. In this chapter we will describe the use of benefit–cost analysis to justify the adoption of a GIS, beginning with a discussion of how the economics of the public sector differ from those of the private sector.

PUBLIC SECTOR ECONOMICS

The discussion of public sector economics in a book whose objective is to provide a theoretical framework for an understanding of managing GIS may seem excessive; however, it is not without purpose. The goal in providing this discussion is two-fold: first, to demonstrate that public sector economics are unique and to describe the differences between public and private sector economics; and second, to illustrate through this demonstration and description the importance of those differences.

Economics in the public sector are different from those in the private sector, a point with which Smith and Tomlinson (1992) agree, and which they have considered in their method for assessing the benefits and costs of GISs. An obvious difference between the two sectors is that the private one requires that an organization turn a profit in order to survive. In the public sector, profit is not a motive; instead, serving the public interest takes precedence. GISs may play an important role in this respect. On the one hand, making the decision to adopt a GIS requires an understanding of the technology as a public good or service. On the other hand, once in place, the GIS may be used to help make other decisions about the provision of public goods and services. It is therefore worthwhile to understand the basics of public sector economics.

Providing goods and services in the public sector can be a complicated undertaking because the nature of public goods means the absence of a pricing mechanism based on typical supply-and-demand functions. Governments normally use taxes as the primary means to raise revenues, then use those revenues to provide a set of goods and services that it thinks match the quantity and mix of goods and services that its citizens need or want. In general, while every member of the citizenry receives some public goods and services, the amount of services that any single individual or family receives is rarely equal to the amount of taxes they pay. The obvious and inevitable result of the taxation–expenditure process is a redistribution of wealth. These redistributive effects, whether they are deliberate or unintentional, may cause conflict among members of the public, either as individuals or as groups. For example, some individuals risk imprisonment by refusing to pay their taxes because they disagree with the way in which the government spends their money. Moreover, lobbyists hover around legislative halls trying to influence the budget process. Both activities reflect conflict over the redistribution of wealth inherent in public budgeting.

Public goods may be described as having a place on a continuum, with "pure" public goods at one end and "impure" public goods at the other. "Pure" public goods are those that are provided to every member of the citizenry and that cannot feasibly be withheld from other members (Samuelson, 1954; Olson, 1971). National defense and local police and fire protection are the classic examples of "pure" public goods. When the federal government sets out to defend the country, the need to cover all parts of the country in order to provide a sound defense results in inclusion of all land and all people.

However, not all public goods are "pure." In the case of an "impure" public good, not only is it possible to exclude some members of the citizenry, but it may in some cases also be desirable, or even necessary, to do so. Oakland (1972) has argued that when an increase in the number of individuals who are served compromises the quality of the public good or service for all users, then rationing or instituting user fees may be indicated. He cites public swimming pools as an example of this phenomenon.

"Impure" public goods are those from which people who do not wish to participate are excluded. Often the government charges a fee for "impure" goods. For example, the federal government charges ranchers a fee to graze their animals on federal land, while local governments charge a fee to those who wish to take a dip in the municipal swimming pool or ride the city bus across town.

A GIS qualifies as a "pure" public good. When a government implements such a system, it does so with the idea that all areas within its jurisdiction will ultimately be included in the database. This ensures that the GIS operates properly. The government does not charge individuals a fee to be included in the database; similarly, all taxpayers (and nontaxpaying residents of the jurisdiction) share in the economic benefits of implementing the GIS.

Perhaps the most vexing aspect of a "pure" public good is the difficulty in assessing its economic (dollar) value. In the private sector, firms manufacture a product or deliver a service and set their price for the good or service at a level that they think potential customers will accept. If the good or service finds few or no buyers, the firm may lower its price. On the other hand, if demand for the product is very high, the firm may raise its price. In short, the willingness of people to pay is the basis for valuing a good or service in the private sector. In the late 1980s and early 1990s, the airline industry would be an example of one sector that has used price breaks to lure flyers back into the skies. "Pure" public goods have no such pricing mechanism. Hence, other logic for valuing public goods must exist.

Early debates about providing and valuing public goods focused on a discussion of taxation and expenditures and may be found in the writings of Adam Smith and John Stuart Mill (Musgrave & Peacock, 1958). Smith suggested that citizens should contribute to the public wealth and fund public goods and services according to their ability to pay. In taking this position, Smith assumed that wealthier citizens gained greater benefits from the state's protection than did the poorer citizens, partly because wealthier people own more property and material goods that are subject to loss in the case of war or criminal activity. By implication, the wealthy have a greater stake in defense and police protection, and, obviously, they also have a greater ability to pay than do middle- and working-class people. The first federal budget to pass in the Clinton Administration (which targeted the richest U.S. citizens for additional taxes) put this theory into practice.

Mill modified the principle of ability to pay by introducing the concept that taxation should inflict equal sacrifice upon all its citizens. This approach would have the benefit of minimizing aggregate sacrifice among the citizens. Moreover, this approach seems to suggest that a progressive taxation system is better than a regressive system. Again, the first Clinton budget is consistent with this philosophy as well.

In later years, other economists began to discuss the expenditure side of the equation. In this respect, the early work of Lindahl (1919/1958) is regarded as a critical innovation in public finance theory, representing the first recognition and acknowledgment of the synchronicity of taxation and expenditures in public finance. The "Lindahl Solution" to the problem of collecting revenues and providing goods and services relies on a Pareto type of optimal approach to taxation and expenditures. Under the terms of this solution, competing groups within a jurisdiction would set the demand for a public good based on their willingness to shoulder the requisite tax burden. Demand would rise to the point where the marginal utility of the good or service would equal price; the point where Group A's marginal utility intersected with Group B's would indicate the only distribution of costs that both groups would find acceptable. The utility principle suggests that there is a direct correlation between the usefulness of a good or service and the willingness of a rational economic actor to pay for it; that is, the more useful an item is, the more a person is willing to pay for it. Simply stated, value derives from use.

Following on the work of Lindahl, Samuelson (1954) also adopted a Pareto type of optimal approach to goods and service provision that relies on utility functions. Implicit in his model is a signaling phenomenon, whereby participants in the system signal (to their fellow citizens) their approval (or disapproval) of various collective goods. On this basis Samuelson hypothesized that "it is in the selfish interest of each person to give *false* signals, to pretend to have less interest in a given collective consumption activity than he really has" (388), hoping to enjoy public goods and services while avoiding the taxes needed to fund the activity. The citizen who acts in this way is characterized as a "free rider." However, the consequence of false signaling in the aggregate, according to Samuelson, is that the jurisdiction will always provide a suboptimal level of public goods. Some might argue that the economics of the Reagan Administration bear this out as, on the one hand, people complained about "big government," while on the other hand they complained that government did not serve them well.

There has been some disagreement about the validity of Samuelson's conclusion. Williams (1966) argued that "spillover" effects (particularly spillover benefits from neighboring jurisdictions) complicate the process by which local governments make decisions on public expenditures, thus producing indeterminate results. In some cases, oversupply results, in other cases, undersupply. Faulhaber (1975) argued that, under certain circumstances, spillover effects can produce Pareto-superior outcomes, which improve the relative positions of all jurisdictions involved and ultimately provide an appropriate level of goods and services.

Another important contribution is Tiebout's (1956) theory of local expenditure, which suggests that providing an appropriate mix and quantity of public goods and services in a given jurisdiction entails a three-step process: (1) forcing citizens to reveal their preferences, (2) satisfying those preferences, and (3) taxing the citizens accordingly. He hypothesized that a major device by which citizens reveal their preferences is to move to a jurisdiction that provides the preferred mix of public goods and services. This aspect of the Tiebout hypothesis has been roundly criticized not only for its neglect of voting as a significant, if not the major, means by which citizens reveal their preferences for public goods and services, but also for ignoring the fact that some people are unable to "vote with their feet" owing either to poverty or to discrimination (Deacon & Shapiro, 1975; Bennett, 1980; Clark, 1981; Whiteman, 1983). Empirically, the flight of both people and businesses from the inner city to the suburbs lends credence to the Tiebout hypothesis.

Further refinements of these models illuminate the complexity that accompanies the revelation of preferences and the development of utility functions for public goods and services. Specifically, the ability of citizens to substitute private goods or income for public goods and services makes it difficult to represent individual as well as aggregate utility functions accurately. Aaron and McGuire (1970) identified two divergent views on household utility functions for public goods: On the one hand, they suggested that the utility of public goods may rise with income, which is consistent with Adam Smith's reasoning that the wealthy individual desires and receives more government protection than do people with lower incomes. On the other hand, the utility of specific public goods may decline with increasing income. For example, Aaron and McGuire pointed out, a swimming pool is far less important to the family that already owns its own pool than it is to a family that does not own one. Using this logic, the individual who can afford to substitute private goods for public goods may prefer to receive fewer public goods as a way to avoid paying taxes to support goods and services that have low utility (thus giving rise to the "forced rider" phenomenon).

Taking this latter argument to its extreme, Maital (1973) argued, requires the government to supply *fewer* public goods (rather than more) in order to promote a more equal distribution of income after taxation and public goods provision. However, the accuracy of Maital's prediction apparently hinges on a specific prioritization of government expenditures for goods and services such that transfer payments and other explicitly redistributive payments take priority over all other payments, leaving until last those payments that primarily benefit wealthy individuals (e.g., defense or police protection). In reality, government budgets are devised through negotiation; and once agreement on individual expenditures has been achieved, no single item has priority except in unusual circumstances. Moreover, defense and police protection are typically top priorities.

There are other criticisms of Aaron and McGuire. Brennan (1976) suggested that under the terms of this model, an equal tax burden for individuals with identical incomes would result in a redistribution of income equivalents from individuals with high demand for public services to those with low demand. Catsambas (1982) also noted the problems that may arise when private goods or services may be substituted for public goods and services, a concern shared by Smith (1981), who noted that self-help and help obtained through social and familial networks are another source of substitutes for public goods. These substitutes are generally more available to people who themselves have greater wealth. In short, assessing individual preferences for public goods is nearly impossible (Hori, 1975; Brennan, 1976; Meerman, 1978).

There is one other aspect of public sector economics that is worthy of mention: the role of externalities. Papageorgiou (1978) defined an "externality" as a negative side effect of an economic activity. The term derives from the fact that private firms, particularly in the past, have been able to "externalize" the unintended negative consequences—and costs—of their activities. For example, the buildup of industry in the United States has resulted in a number of toxic waste dumps, as manufacturers discarded their waste material where convenience dictated. To be fair, some of these materials were not known to be toxic at the time they were dumped. Now that we do know the potential for calamity created by such dumping, governments at all levels have placed stricter controls on toxic waste. Furthermore, the federal government has in place a program (Superfund) that is intended to pay for the cleanup of toxic waste sites. By all rights, the individuals and firms responsible for creating the sites should be liable for the cleanup. This has not happened; instead, some firms have succeeded in shifting or "externalizing" the cost of cleaning up their manufacturing waste. Someone else is paying.

Externalities may be desirable, as Williams' (1966) "spillover" effects (mentioned above) imply. For example, the commuter rail service that serves Chicago and its suburbs stops in the Hyde Park neighborhood, a few miles south of downtown, primarily because it is the terminus of the trips of suburban train riders who may be employed at the Museum of Science and Industry or at the University of Chicago. This has a favorable outcome for Hyde Park residents, who can get on the train at this point for the quickest, safest, most comfortable trip downtown. Faulhaber (1975) argued that situations such as these improve the relative positions of all jurisdictions involved and ultimately provide an appropriate level of goods and services.

According to Smith and Tomlinson (1992), governments cannot externalize their costs. Their missions are so broad in nature, and their clientele so all-inclusive, that externalities are, for all practical purposes, nonexistent. For example, firms that dumped toxic waste that eventually caused illness or death suffered no economic loss because of that illness or death. On the other hand, governments recognize that avoidable illness and death are an economic burden, causing a rise in medical expenditures, leaving dependents impoverished, and removing productive workers from the tax rolls. Governments have nowhere to shift these burdens.

While the problems of public sector economics are significant, they do not prohibit an organization from making a reasonably objective and sound decision regarding the implementation of a GIS. Indeed, while there is not a huge volume of literature relating to this type of decision, what does exist consistently points to the benefit–cost analysis (or cost–benefit analysis) as the best means of evaluating the efficacy of implementing a GIS.

BENEFIT–COST ANALYSIS

The use of benefit–cost analysis as a justification for adopting a GIS is well established in the literature on GISs (e.g., Dickinson & Calkins, 1988, 1990; Wilcox, 1990; Aronoff, 1989; Huxhold, 1991; Smith & Tomlinson, 1992). The benefit–cost analysis is commonly used in both the public and private sectors to determine the economic wisdom of making large expenditures. The basic idea behind benefit–cost analysis is to calculate the total value of the benefits of the expenditure versus the costs of the expenditure. A ratio of one to one or greater justifies the expenditure. In the case of familiar expenditures, the calculation is relatively simple because the full range of costs and benefits are well known. However, calculating the benefit–cost ratio is more complicated and uncertain where the expenditure is unfamiliar. According to Dickinson and Calkins (1988), this is the case for GISs.

In evaluating the wisdom of purchasing a GIS, both the benefits and the costs of implementation may be difficult to assess because of the uncertainty surrounding them. It is well known and generally accepted that the costs of implementing a GIS extend beyond the purchase of hardware and software. For example, the assembling and maintaining of data and the training of staff are two areas that will require expenditures after the initial purchase of a GIS. The exact dollar amount of these additional costs is often difficult to know ahead of time. However, as Smith and Tomlinson (1992) noted, "The costs [associated with implementing a GIS] are loaded heavily in the early period whereas the benefits increase . . . and then remain constant" (254). Similarly, because the GIS is still a relatively new technology, the full benefits of the technology may be either over- or underestimated at the time of implementation. After the system has been in use for some time, the organization will begin to realize the system's full value. For example, the organization may find new and unexpected uses for the technology that will enable it to increase its productivity or cut its costs.

The uncertainties surrounding the calculation of benefits and costs of implementing a GIS have been the subject of discussion by several authors. Both Aronoff (1989) and Huxhold (1991) described some of the difficulties in accurately assessing the benefits and costs of implementing a GIS. Huxhold suggested that there are three major categories of benefits that a benefit–cost analysis should examine: (1) cost reduction, (2) cost avoidance, and (3) increased revenue (244) (see Table 6.1).

TABLE 6.1. Benefits from GIS Adoption

Huxhold	Aronoff
Cost reduction	Increased efficiency
Cost avoidance	New nonmarketable services
Increased revenue	New marketable services
	Better decisions
	Intangible benefits

Note. Adapted from Huxhold (1991) and Aronoff (1989).

Huxhold defined cost reduction as "the decrease in operating expenses of the organization, primarily caused by a savings in time by operating personnel performing their tasks more efficiently" (244). Cost reductions generally accrue because of the improved productivity of staff members responsible for the tasks performed using the GIS. Drafting, map updating, and so on are tasks that may be included in these reductions. In addition, an organization might specifically use its GIS to improve the efficiency of its operations. For example, according to Huxhold, a GIS may result in time savings by optimizing service-district boundaries, making it possible to use existing staff more efficiently, and, therefore, reduce the number of service delivery staff (244).

Huxhold defined "cost avoidance" as the "prevention of rising costs in the future caused by projected increases in workload" per staff member (246). This benefit is consistent with, and more or less an extension of, the first benefit, suggesting that once a GIS becomes part of an organization's equipment, it may help to optimize the performance of a variety of current and future tasks. According to Huxhold, this improvement in performance may make it unnecessary to hire new employees, or at least postpone such hiring by making the best use of existing employees (246).

Finally, Huxhold suggested that "a GIS can increase revenues . . . by selling data and maps, increasing property-tax collections, and improving the quality of data used to apply for state and federal grants" (245). The rationalization of tasks that the GIS effects does indeed bode well for the increase in tax collections and the improvement in data quality. However, Dansby (1991) suggested that there may be legal impediments to the sale of such products in the public sector, depending on state and local regulations on copyrights and freedom of information.

Selling Data

According to Dansby (1991), whether it is legal to sell GIS products created in the public sector depends on the applicability of two conflicting laws in each specific case: freedom of information (open records) and copyright protection (100). State and local governments have the same copyright protection that is available to individuals and other entities; the federal government, however, does not. Dansby suggested that, in general, GIS databases developed by state and local governments could be considered as original works of authorship, which, therefore, would be protected by copyright. In cases where geographic databases contain material that originated with third-party authors, the third-parties (not the government agency) would own the copyright to the data. The ownership of a copyright is consistent with selling products for cost-recovery purposes. Therefore, governments that own the copyright to the information in their GIS databases have met the first condition for selling the information.

Dansby pointed out that in potential conflict with copyright laws, open-records laws frequently include clauses that prohibit the sale of the records for any more than a nominal fee (for example, the cost of reproduction) (101). Dansby argued that the validity of using copyright protection as a rationale to charge fees for GIS products or other government data "depends on the wording of each state's open records laws. . . . Nothing can prevent cost recovery to help finance GIS development in those states whose open records laws are narrowly drawn to allow inspection only. Other states would require that exceptions to the open records laws be passed or that the laws be amended" (101). In short, the potential benefits of selling GIS products are not available to every jurisdiction that employs a GIS, and, therefore, should not be included automatically in every benefit–cost analysis.

Moreover, there is an ethical argument against selling government data at a profit, based on concerns about charging the public twice for the same information. In most cases, taxes fund government activities, including the development of databases and GISs. These taxes come from people living in the jurisdiction and take the form of income taxes, sales taxes, property taxes, or fees for use. The argument that it is best to charge merely for the cost of reproduction of information rests on the acknowledgment that the citizens have already paid to gather and develop the information. Any profit taking would appear to be double charging, and may alienate the citizenry.

On the other hand, making available at a nominal cost information that a purchaser may use in a secondary profit-making activity
poses additional ethical problems. Is this an appropriate use of data
collected at public expense? And isn't the purchaser also a member of
the public? Royalties, as opposed to outright purchase, have been
suggested as a means to address this concern.

An additional concern about selling data is liability for outcomes
of unintended uses or deliberate abuses of the data. Once data and
information find their way into the hands of purchasers, the originators
no longer have control. Unintended and inappropriate use, as well as
deliberate abuse, of data and information may result in negative
consequences. Who is liable? The purchaser only? Or is the government unit that sold the data also liable? With potentially large cash
settlements at risk, this is not a trivial matter.

The problems attending the potential benefit of selling data
suggest that there is great uncertainty associated with including this
"benefit" in any benefit–cost analysis. Many details of such sales
remain unresolved. The passage of time and the accumulation of
experience will lead to the drafting of appropriate legislation and the
development of case law to address these problems. Until then, it is wise
to use extreme caution in including sale of data and information as a
"benefit."

Additional Benefits

There are other ways to approach benefit–cost analysis. Aronoff
(1989) identified five separate types of benefits that could result from
the adoption of a GIS: (1) increased efficiency, (2) new nonmarketable services, (3) new marketable services, (4) better decisions, and (5)
intangible benefits (260–261) (see Table 6.1). Aronoff's ideas of
increased efficiency and new marketable services correspond to
Huxhold's notions of cost reduction and increased revenue, respectively.

According to Aronoff, new nonmarketable services are "useful
products and services that were previously unavailable" and will be
used within the organization (260). Aronoff noted that the organization will anticipate some of these GIS benefits. Some of these benefits,
however, will not be apparent until after the GIS is up and running.
Therefore, it will be difficult to assess this benefit of new nonmarketa-

ble services and include it in the benefit–cost analysis in order to determine the wisdom of adopting a GIS.

Aronoff also suggested that the adoption of a GIS will produce "better decisions" (261). He speculated that this will occur because "more accurate information and faster and more flexible analysis capabilities can improve the decision-making process itself" (261). Again, determining the economic value of these "better decisions" is difficult, at best.

Finally, Aronoff said that adopting a GIS may bring intangible benefits to an organization. Smith and Tomlinson (1992) defined "intangibles" as "not as much a separate category of benefits as they are a class of benefits that is more difficult to quantify" (249). According to Aronoff, these benefits might include such things as better internal communication in the organization, improved morale, and a better public image (261). By definition, placing an economic value on these intangible benefits is not possible. Organizations may, however, describe these potential benefits in text accompanying the benefit–cost analysis.

The handling of externalities in the benefit–cost analysis is another matter that merits mention. In the case of governments that are performing a benefit–cost analysis as a prelude to their implementation of a GIS, Smith and Tomlinson (1992) recommended incorporating "all benefits . . . of the analysis whether or not they accrue to the potential GIS purchaser or to the departments that will use the information products" (250). According to Smith and Tomlinson, among the nongovernment groups that may realistically expect to benefit from the implementation of a GIS are taxpayers, private firms, and special service districts (250).

Payback Period

Another variation of the benefit–cost analysis is the calculation of the "payback period," discussed by Huxhold (1991, 249). This calculation involves dividing the total cost of implementing a GIS by the estimated value of annual benefits of using the system. The resulting figure tells how many years it would take to accumulate enough benefits to pay for the cost of the system. The benefits may include any or all of those described earlier in this chapter. Not surprisingly, this calculation is fraught with the same difficulties apparent in typical benefit–cost analyses.

Discounting

One last refinement of the benefit–cost analysis is the practice of discounting, described by Smith and Tomlinson (1992). The idea behind discounting is to deflate the data relating to costs and benefits in order to remove the effects of inflation. According to Smith and Tomlinson, this practice takes into account "that benefits realized in the year 2000 are less socially valuable than the same dollar amount of benefit realized today" (256). In short, discounting provides a mechanism for operationalizing the old saw that money doesn't go as far as it used to. Discounting is important for providing an accurate assessment of the value of implementing a GIS because of (1) the multiyear life expectancy of a GIS, and (2) the resulting fact that GIS costs and benefits are also spread over multiple years.

The benefit–cost analysis is the preferred method for justifying the implementation of a GIS. However, an organization must use caution when performing such an analysis to determine if it is wise to adopt a GIS. The GIS must be assessed in comparison with existing practices and technologies used in the organization. In addition, the organization must consider other new means of performing the same tasks. Making these comparisons requires the organization to perform a separate benefit–cost analysis for each alternative under consideration. The alternative with the highest ratio of benefits to costs is the most efficient one.

The purpose of this discussion of benefit–cost analysis has been to briefly describe the method and to point out its advantages and disadvantages. The method's primary advantage is that it provides an economic rationale for an organization's decision to adopt a GIS. This can be very persuasive. The disadvantage of using benefit–cost analysis is that it is very difficult to assess or quantify many of the benefits and costs that are needed to perform the calculation. Organizations that wish to do such an analysis may find it helpful to follow the example of other organizations that have undertaken this task previously and, obviously, to be cautious in making their own assessments.

CONCLUSIONS

The public sector has been a fertile field for the adoption of GISs. The fact that the public sector produces and uses a large amount of geographically referenced data is a key reason why this sector has embraced GISs. Typically, organizations considering GISs use benefit–cost analyses to help them decide whether or not to purchase such systems. However, although these analyses are useful, they cannot speak to the value of the GIS as a public good.

Public organizations must also concern themselves with whether GISs are an appropriate public good within the context of their own jurisdictions. This will depend on the specific wants and needs of the relevant citizenry. A basic understanding of the rudiments of public sector economics will provide a foundation for making these judgments.

❖

Sharing Geographic Information across Organization Boundaries

As development of GISs progresses and technical problems are overcome, new problems arise. One of the frequently mentioned problems is that of the need for information and databases that may be housed in several organizations. In this chapter we will explore the organizational and managerial roots of difficulties in sharing databases. Historically, a combination of an organization's structure and operations have resulted in the fragmentation of work projects into individual tasks and growing powerlessness among manual and clerical workers. In contrast, such fragmentation of tasks tends to increase the power of knowledge workers and their organizations, since such workers and organizations often control information that is valuable in that it is unique and indispensable. In this chapter we will identify three means by which alliances may be formed and information may be shared: appeals to professionalism, coercion, and bargaining. We will propose a theory based on the relative power of the participants to predict which of these three strategies will be used. Finally, we will propose a three-stage conceptual model that delineates (1) a set of facilitators of information sharing in an organization, (2) the process of information sharing, and (3) consequences of enhanced information sharing.

With the steady rise in acceptance of GISs in recent years, there is general agreement that the success of organizations in both the public and private sectors can be greatly enhanced by the open exchange of geographic information across organization boundaries. Certainly, from a technical perspective, data is now far easier to share owing to its electronic form. However, the rapid increase in the number of organizations adopting GIS technology has belied the fact that between and among organizations there has been a general inability, and often unwillingness, to share data and information across organization boundaries. The waste caused by duplication of effort—which is due largely to lack of information exchange among local, state, and federal governments, and the private sector—remains a significant impediment to the more effective and efficient use of GIS throughout society.

A favorable development has been the recent initiative sponsored by the National Center for Geographic Information and Analysis (NCGIA), which has sought to call attention to the need for increased sharing of geographic data sets as a first step toward improving coordination among GIS-using agencies and organizations. Developing in organizations methodologies that provide corporate and personal incentives for internal and external sharing of geographic information in tandem with creating appropriate technical standards will help facilitate sharing of information among those who have acquired the technology. Such methods will serve also to advance the acceptance and use of geographic information within organizations that have not, to date, adopted GISs in their operations.

In this chapter we will seek to examine some of the roots of the problem of gaining cooperation across organization boundaries in sharing geographic information. We will begin with a discussion of specific aspects of practices in organizations by managers and the role of these practices in the fragmentation of information gathering and data storage critical to the development of GISs. In particular, we will discuss the importance of Taylorism, on the one hand, and of Weberian concepts of expertise and client groups, on the other. In suggesting a means to reintegrate the data, we will examine the literature on the formation of alliances, which may be understood as the organization of disparate groups for the purpose of achieving a specific goal. As a result, we suggest a hypothesis to predict under what circumstances organizations will employ various strategies to effect information sharing, based on their power relative to that of other organizations involved. Finally, in this chapter, we will also propose a conceptual framework that addresses the influence of a set of antecedent constructs (accessibility, incentives, superordinate goals, bureaucratized and formalized rules and procedures, the quality of exchange relationships among organizations, and resource munificence/scarcity) on the attainment of both cooperation among organizations and optimal use of GIS information.

ROOTS OF THE
DATA-SHARING IMPERATIVE

Early organization theorists often argued for the "rationalization" of specific tasks in the workplace through restructuring production processes and encouraging wide-scale job redesign. Historically, this combination of the structure of organizations and operations con-straints has resulted in the fragmentation of work projects into individual (often repetitive) tasks, resulting in growing powerlessness among manual and clerical workers. By contrast, the fragmentation of tasks has had quite a different result among knowledge workers, particularly those with technical expertise in crucial areas. Knowledge workers may possess information that is extremely valuable because it is both unique and indispensable to particular applications. Under such circumstances, the developer or owner of unique and indispensable information possesses a key element of control.

We contend that there are three key means by which interagency alliances are formed and by which information can be shared: (1) appeals to professionalism, (2) coercion, and (3) bargaining. The strategy employed to achieve information sharing, we suggest, will depend primarily on the relative power of the organizations involved.

Taylorism and the Principles of
Scientific Management

An influential and seminal work in the field of management is Frederick Winslow Taylor's *Principles of Scientific Management* origi-nally published in 1911. Beginning with the premise that increased productivity brings lower unit costs—which are advantageous to em-ployer, employee, and customer (or client) alike—Taylor argued that "the best management is a true science, resting upon clearly defined laws, rules, and principles, as a foundation" (7). Furthermore, good management lies not with particular individuals but in basic knowledge and skills that can be learned. Taylor coined the phrase "scientific management" to describe his theory of management.

The principle of scientific management calls for the development of a science to replace the "rule of thumb" knowledge of workers. The focus of scientific management is the task. As Taylor (1911) himself noted,

Perhaps the most prominent single element in modern scientific management is the task idea. The work of every workman is fully planned out by the management at least one day in advance, and each man receives in most cases complete written instructions, describing in detail the task which he is to accomplish, as well as the means to be used in doing the work. And the work planned in advance in this way constitutes a task which is to be solved, as explained above, not by the workman alone, but in almost all cases by the joint effort of the workman and the management. This task specifies not only what is to be done but how it is to be done and the exact time allowed for doing it. (39)

According to Taylor, within this general notion of the task idea is an explicit description of the "subdivision of the labor; each act of each mechanic, for example, should be preceded by various preparatory acts done by other men. And all of this involves, as we have said, 'an almost equal division of the responsibility and the work between the management and the workman' " (38).

Taylor's *Principles of Scientific Management* was widely accepted, both in theory and in practice, in Europe and the United States in the early part of this century. Urry (1986) has suggested, however, that the adoption of scientific management occurred more rapidly in the United States than in Europe, partly because "technological changes [had] outstripped the capacity of craftsmen trained in traditional techniques to organize production in the way they had in the past" (46). Indeed, the implementation of scientific management techniques in the early 1900s was greatest in new industries, such as the automobile industry. One of the major contributions of Henry Ford, for example, was the implementation of the assembly line, which is wholly consistent with and may be perceived as an outgrowth of the principles of scientific management.

The implementation of scientific management has not been without consequences. One important consequence has been the growth in the power of management even as there has been a diminution of the power of labor at the lowest levels of the hierarchy (Lipietz, 1986; Sayer, 1986; Urry, 1986). At the same time, however, Urry has suggested, there has been an increase "in the numbers and influence of industrial engineers" and that their relationship with management has become "increasingly symbiotic" (46). While Taylor's *Principles of Scientific Management* help us understand the origins of the fragmentation of tasks, Weber's theory of bureaucracy helps explain the growth in power among engineers and other professionals in spite of managerial tasking edicts.

Weberian Concepts of
Expertise and Professionalism

Since it was first translated into English in 1946, Weber's theory of bureaucracy has become the keystone of organization theory in the United States. There are some important consistencies between Taylor's *Principles of Scientific Management* and Weber's theory of bureaucracy, such as office hierarchy and graded authority resulting in a finely ordered system in which higher offices supervise lower ones. Another similarity is that the management of the office in a bureaucracy follows general rules that are more or less stable, more or less exhaustive, and that can be learned (Weber, 1946).

In sharp contrast to the manual worker described by Taylor, Weber's bureaucrat is frequently trained or educated outside of the institution where he or she is employed, often at a college or university. This specialized training is often coupled with some sort of certification, such as the earning of academic diplomas (the Ph.D., e.g.) or the successful passing of professional examinations (the bar exam, for example). This coupling provides objective evidence that the individual has mastered specific professional skills.

Over time, the bureaucracy may come to occupy a position of considerable power within its jurisdictional sphere based on the expertise of its professional staff (bureaucrats). In the case of the public bureaucracy, the legislature (or some other body of elected officials) is—nominally at least—the "master of the bureaucracy." Yet the master rarely has the degree of expertise that the servant—bureaucracy—possesses. In practice, this phenomenon means that the political master frequently plays the role of "dilettante" to the bureaucracy's "expert." Particularly in the case of highly technical fields, the master often must defer to its more knowledgeable servant.

The related source of the bureaucracy's power is the professionalism of bureaucrats and their ability to guard their professional knowledge from outsiders. Each profession has a body of knowledge that is uniquely its own. In order to become a member of the profession, the hopeful candidate must go through a more or less well-defined rite of passage that ordinarily culminates in some sort of certification process. For example, if a person wishes to become a lawyer, he or she attends an accredited law school, graduates, takes the bar examination, and—provided he or she passes the exam—is admitted into the practice of law.

The profession that can closely guard its body of knowledge

improves its staying power. Technical professions are often very successful in this objective because they have learned to guard carefully a relatively small but highly complex body of knowledge that is theirs. While limiting entry into the field, certification serves a valuable purpose by helping to preserve the technical credibility of the group within its field of expertise.

In contrast to the emphasis that Taylor placed on management and control, Weber recognized the power inherent in professional expertise. In contrast to manual workers, the expertise of professionals limits the absolute control of management. In fact, professionals in bureaucracies discover in some instances that it is to their advantage to use their expertise to control a very small part of a larger project—especially if that small part is crucial. Moreover, as recognized experts in their fields, bureaucrats are in a position to redefine their tasks, emphasizing the technical details and the expertise required. Unless management possesses equivalent or greater expertise, bureaucrat–experts can maintain significant control over the work itself.

Fragmentation of Tasks

Whereas scientific management promoted the subdivision of tasks within individual organizations, bureaucracy tends to subdivide tasks and delegate them to separate organizations or separate departments. There are two reasons for the development of specialized bureaucracies, both of which are consistent with Weberian theory: expertise and client groups.

The notion of expertise as a requirement for public employment has become widely accepted and implemented, as evidenced by the slow demise of political patronage and its replacement with the Civil Service system as the predominant system for hiring government employees at all levels. The development of increasingly specialized academic programs of study and professional school programs reinforces this phenomenon by providing a ready supply of trained professionals. In some organizations, professional hegemony assures that only individuals with a specific degree, and sometimes from a limited group of institutions, find employment in a particular organization. This may occur intentionally as in the practice at General Motors Corporation of hiring engineers trained at its affiliate school, General Motors Institute. Or it may occur unintentionally or coincidentally as a by-product of "old-boy networks," for example.

The expertise of individuals in the organization becomes part and parcel of the organization itself. As noted earlier, this gives the organization tremendous authority within its area of expertise. The notion of expertise is so well accepted, in fact, that even the U.S. Supreme Court is loath to overturn substantive agency decisions suggested by technical experts (Yellin, 1981).

The second reason for the development of specialized bureaucracies is the linking of organizations with specific client groups. Each organization that comes into being has a specific mission or purpose. This mission or purpose is generally linked with some client group. For example, the U.S. Department of Housing and Urban Development was created to address the needs of the urban poor, evidence from the Reagan Administration of administrative malfeasance notwithstanding. Similarly, the Pentagon has a smaller number of very powerful clients within the U.S. military–industrial complex.

As increasingly specialized client needs come to light, new organizations arise. For example, the U.S. Department of Health, Education, and Welfare was reorganized by President Carter in 1980 into the Department of Health and Human Services and into the Department of Education to better serve these very different client groups. The creation of new agencies also benefits the bureaucrats who possess the necessary credentials to assume positions within them. Moreover, the same interest that leads the bureaucrat to earn credentials in a specific field may also provide him or her with the motivation to perform to the best of his or her ability within an organization. Opportunities for career advancement also serve as motivation for the bureaucrat. In short, it is difficult to untangle the ways in which expertise and client groups contribute to the development of specialized bureaucracies, although it is clear that both play a role.

Fragmentation of Data

The combination of the subdivision of tasks and the development of increasingly specialized bureaucracies has as one of its consequences and corollaries the fragmentation of data. Even as government functions have been split among a variety of agencies, the agencies charged with specific functions have responded to their mandates by gathering information and data essential for the performance of their work.

It has become a truism that "knowledge is power." If this is so, then what of the components of knowledge—data and information? Again, the concept of professionalism is relevant. It is not uncommon for organizations to protect their knowledge, information, and databases as a means of exerting their professional hegemony and of demonstrating their expertise, which, in turn, facilitates the protection and expansion of professional turf. This can create a snowball effect, as those who control information use it to advance their knowledge, thus improving their expertise, which can fuel the cycle.

As long as an organization maintains control of valuable information, it can remain dominant and protect, or even improve, its professional position. Should the organization lose control, or should the information lose its value, the organization's position will weaken. The organization, therefore, has a strong incentive to maintain control of any information it possesses and to use its control wisely.

The fragmentation of information and data is far different in its effects from the fragmentation of tasks. Whoever possesses valuable information needed for the completion of a larger task is valuable. The knowledge worker may possess unique and necessary information, and in this way may resemble more closely the last piece of a jigsaw puzzle. Even though the piece may be one of a thousand pieces, its uniqueness makes it irreplaceable. By contrast, in many cases the labor of the manual worker has been simplified to the point where he or she may be readily replaced, much as we purchase a new air filter for our car.

If we agree that the possession of information serves as a source of control for individuals and organizations, then we are faced with questions about the ways in which organizations can be induced to relinquish this control. The literature on alliances may offer some insight into resolution of this problem.

INFORMATION-SHARING ALLIANCES

There are several bases for the formation of alliances, or intergovernmental systems, as they are sometimes called. Olson and Zeckhauser (1969) and McGuire (1974) suggested that even when people (or organizations) are devoid of feelings (either positive or negative) toward one another, they may find that it is in their interest to organize for the purpose of providing collective goods. Among the bases for the formation of alliances are professionalism, coercion, and bargaining.

Appeals to professionalism may sometimes be a motivating factor in the development of alliances that can facilitate interagency information sharing. Milward (1982) and Keller (1984) noted the importance of functional interests and professionalism in establishing intergovernmental systems. Gage (1984) argued for the importance of understanding such networks as "instruments for establishing and maintaining political networks to accomplish policy objectives" (136). In some cases, professionals may respond to a sense of professionalism, putting aside interagency rivalries to pursue a common goal. Such short-term sublimation of individual and agency goals to address a larger picture can sometimes result in long-term benefits aside from the achievement of a specific common goal. Such joint ventures may stimulate employment opportunities in the field and foster consistent support for the agencies in specific policy areas.

In some instances, the development and maintenance of alliances will be far more difficult, depending on the negotiation of an acceptable exchange among the parties involved. As Weber asserted, "Rational exchange is only possible when both parties expect to profit from it or when one is under compulsion because of his own need or the other's economic power" (Weber, 1968, 73).

In general, intergovernmental systems (or networks, as they are sometimes called) are characterized by an uneven distribution of power (Lindahl, 1919/1958; Stern, 1979; Milward, 1982; Keller, 1984). According to Milward, within this environment of uneven power distribution, it is not unusual for factions to compete for power to assure that their goals are ultimately adopted and implemented by the intergovernmental system as a whole (470). The result is an ongoing search for equilibrium among the members of the system (see Milward, 1982; Keller, 1984). Equilibrium is not a static condition, but a process outcome that the members of the system achieve through their efforts to gain power. Inasmuch as the quest for power is ongoing, conflict arises as the equilibrium point for the system as a whole changes.

While it seems to be inevitable, conflict should not automatically be viewed as a negative element (Pondy, 1967; Buntz & Radin, 1983). Rather, conflict is the means by which the intergovernmental network achieves an equilibrium of power. In some instances, conflict can produce negative effects; in others, it serves an important integrative function within the network (North, Koch, & Zinnes, 1960; see also Pondy, 1967). Keller (1984) noted that as a means of improving their power position within the network, members of the network may attempt to link their own organization missions with the values held by powerful external groups. Pondy (1967) shares this view: "A major element in the strategy in strategic bargaining is that of attitudinal structuring, whereby each party attempts to secure the moral backing of relevant third parties" (313).

A THEORY OF INFORMATION-SHARING STRATEGIES

The literature on alliances suggests three separate ways in which interagency alliances occur: (1) appeals to professionalism, (2) coercion, and (3) bargaining. Appeals to professionalism may in some cases represent an appeal to somewhat altruistic, noble values. In other instances, such appeals may reflect crass self-interest on the professional level. An appeal to professionalism has as one of its advantages its very low cost (i.e., "talk is cheap"). It is therefore readily available to any organization.

The second means by which interagency alliances occur is through coercion. In some instances, coercion comes by way of controls placed on a government by some more powerful level of government. For example, physical development projects of a certain size that are proposed as federally funded efforts are subject to the terms and conditions of the National Environmental Policy Act of 1969. Similarly, state governments may have the authority to require specific information of local governments.

In interagency networks where the power structure is less well defined, or where there is minimal difference in the power of the various agencies, coercion may be impossible. In these instances, bargaining appears to be the most likely means of achieving an agreement on the information.

Within the basic concept of bargaining, organizations have a variety of resources at their disposal. In some instances, information swaps may be possible. Some organizations may have the economic resources to purchase information from other agencies or to provide some other considerations.

Two factors stand out as central to achieving agreements on information sharing: (1) the value of the information to the negotiating agencies and (2) the interagency power structure. Assessing the value of information continues to be a nagging problem. Information does not have value in and of itself, but rather its value is related to its utility to its potential users. One clear indication of the value of information is the price it commands in the market. The very existence of information brokerage firms provides evidence that market methods of valuation do occur. However, assessing the value of information held in public or private sectors remains an inexact art, at best.

We argue that it is possible to identify which of the three types of resolution will occur—appeal to professionalism, bargaining, or coercion—on the basis of the balance of an organization's power. Figure 7.1 identifies under which power structures each of the types of resolution will obtain.

This model assumes that agencies will seek the least-cost resolution. Within this model, the two least-cost resolutions are coercion and appeals to professionalism. Coercion, however, is available only to organizations that possess the power or authority to pursue it. Appeals to professionalism are available to everyone.

Where the balance of power favors the seeker of information, that organization may exert its authority and demand the information of the weaker owner of the information. Where the balance of power favors the owner of the information, the weaker seeker of the information has neither the authority at its disposal to demand the information nor the power needed to enter into a bargaining situation. When the seeker of information is relatively weak, it must rely on appeals to altruistic notions of professionalism and the public good. Bargaining can occur only when both the owner and the seeker of

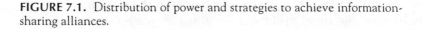

		Owner of Information	
		Powerful	*Powerless*
Seeker of Information	*Powerful*	Bargaining	Coercion
	Powerless	Appeal to professionalism	Bargaining

FIGURE 7.1. Distribution of power and strategies to achieve information-sharing alliances.

information possess roughly equivalent power, although we would suggest that it makes little difference if both are relatively powerful or relatively weak.

The presence of complicating factors, such as the value of the information and the relative power of the agencies involved, gives rise to uncertainty about the resolution of specific cases. Indeed, the relative value of the information in question to the agencies involved is likely to become part of the power equation. Again, because of the difficulty of accurately assessing the value of information, this contribution of the value of information to the relative power of the agencies is unknown. Empirical research is needed to ascertain the validity of the model developed as Figure 7.1 in this chapter. Actual case studies of information sharing will be needed to gather this information. Case studies should yield valuable information about the nature of the interagency bargains agencies adopt to make possible the sharing of information needed for the larger-scale GISs.

ANTECEDENTS AND CONSEQUENCES OF INFORMATION SHARING

In addition to examining some of the means by which interagency alliances are developed, it is our goal to propose a research framework for the study of facilitators of information sharing among organizations. Within the organization-theory literature, the area of interorganization cooperation is one that has seen far less research to date than intraorganization cooperation. Further, little research has examined the types of factors that can have a positive effect on convincing organizations to share information. Our goal is to offer a framework for understanding information sharing based on the expected effects of a variety of factors in organizations regarding cooperation.

In attempting to address the issue of GIS information sharing among organizations, it is first necessary to provide a context within which such information is often needed. We propose three distinct contexts requiring interorganization GIS information sharing. The first information sharing context is situation specific, or project-driven. In this situation, two or more agencies come together to work on a common problem that both parties need to address. The two organizations bring their own data and expertise to the table and share information with each other in an effort to successfully solve the problem. For example, hazardous waste disposal is an environmental problem that may require the cooperation of several federal agencies with overlapping responsibilities. In order to develop a comprehensive and effective solution to a hazardous waste problem, several agencies come together to share GIS information that they each possess in order that all parties can contribute to an optimal solution. As this example illustrates, the problems associated within this context are usually nonroutine and nonrecurring, requiring that they each be addressed on a case-by-case basis with their own distinct solutions.

The second information-sharing context is one in which different agencies may be addressing completely different problems but have a need for very similar information. Because they have needs for the same information, organizations develop procedures by which they can regularly share and exchange information with each other. For example, perhaps the Bureau of Indian Affairs and the Bureau of Land Management determine that they have overlapping needs for information about the same federal lands. This need for common information encourages collaborative (sharing) behavior. Before they begin to exchange information, an analysis is performed to determine the procedures (both technical and administrative) by which these two bureaus can most effectively exchange GIS information. In this context, often the first step in the sharing process is to determine some sort of protocol regarding mutual access to either party's information. The focus is on the needs of the cooperating parties, and meeting those needs might result in either a case-by-case approach or a long-term data-sharing protocol. However, the protocol established to meet the needs of the participating parties may be highly inappropriate for effective sharing with additional parties or the broader community.

The third context for information sharing among organizations is based on developing standardized or generalizable patterns of exchange. In this context, information is readily accessible to all parties and can be accessed in useful forms from a central location, such as a data bank, or from networked or decentralized sources. Organizations simply engage in a routine sharing process through the central storage facility or network of decentralized facilities when they need information.

While all three contexts for information sharing have validity, for purposes of illustrating our framework, we have chosen to focus on the first context for information sharing (i.e., that of organizations working together in an ad hoc manner to solve a specific problem). Our primary reason for selecting this case lies in its representation of a common aspect of the geographic information sharing problem. Readily evident and observable are the many situations that require two or more agencies to exchange GIS information in order to solve a problem that the agencies have a common interest in. As a result, in focusing on this more popular context, we are addressing concerns that are immediate and compelling to a wide range of individuals charged with the task of attempting to develop better methods for information sharing. Further, as a result of gaining additional knowledge into the facilitators of organization information sharing, practitioners and researchers will likely be in a better position to gain insights for those investigating issues of information sharing within the other longer-term contexts articulated above. So many technological and infra-structure issues must be addressed in tandem with the interorganiza-tion issues that any research program will necessarily need to be active and malleable over time.

This section of the chapter focuses on a three-stage conceptual model that identifies both antecedents and consequences of interor-ganization cooperation in sharing GIS information and technologies. Research in the areas of organization power and political behavior (Pfeffer & Salancik, 1978), channel relationships (Dwyer & Oh, 1987), negotiation (Wall, 1985), and intergroup conflict (Thomas, 1976; Walton & Dutton, 1969) have sought, as one of their goals, an increased understanding of the factors by which improved cooperation can occur. While this research has led to an enhanced understanding of cooperation from an organization-theory perspective, an analysis of the factors that can lead to greater sharing has not been attempted in organization environments with a focus on the dynamics of sharing geographic information. Further, the relation between GIS informa-tion sharing, on the one hand, and organization effectiveness and efficiency, system satisfaction, and decision making, on the other, has not received attention to date.

Previous research on cooperation and information sharing has focused almost exclusively on intraorganization collaboration—that is, attempting to better understand how different functional groups within a single organization can develop more cooperative relation-ships. This research stream—best represented by the work of Lawrence and Lorsch (1967); Gupta, Raj, and Wilemon (1986); and Souder (1981, 1988)—empirically examines relationships within an organiza-tion, usually between specific functional groups, and suggests factors that are important in fostering these relationships. The findings deal with the effect of interdependence, or resource dependency, and coordinating mechanisms (i.e., formalized rules and procedures) on cross-functional interaction. In addition, the research results suggest that the similarity of functional departments—as far as duties and objectives are concerned—positively influences the amount and type of communication between the departments. Finally, this research has been very useful in offering prescriptive advice on methods to better facilitate information sharing among different functional departments.

What has been missing from the literature are systematic attempts to develop a framework of antecedent variables that can improve the likelihood of creating positive, collaborative relationships *between* organizations. Specifically, little is known, for instance, about the reasons for governmental agencies and other GIS-using organizations sharing or not sharing GIS-related information. There is strong evidence to suggest that considerable duplication of effort occurs across organizations because of a basic lack of will to cooperate. Although the technical ability to share geographic information might be readily accessible or achievable, the incentives for an organization or a person to share are insufficient to overcome the impediments. In other instances, data and information are held closely as sources of control and power, even within government offices. In these instances, there is often outright unwillingness to share this "proprietary" information.

Little is understood about the factors that can act as facilitators, or antecedents, of information exchange among organizations. This section draws on literature from a number of sources, including organization theory, intergroup dynamics, exchange theory, and political-economy research, in order to posit a model of interorganizational information sharing.

A CONCEPTUAL FRAMEWORK

A variety of factors can act as facilitators or inhibitors of information sharing and cooperation among organizations. These factors range from such individual variables as the personalities of group members, use of information as a source of personal power, "turf" battles, interpersonal relations, and training and skills (Johnson, 1975; Kelly & Stahelski, 1970; Pavett & Lau, 1983; Pfeffer & Salancik, 1978; Schmidt & Tannenbaum, 1960) to such organization variables as political concerns, reward systems, and cultural norms (Lawrence & Lorsch, 1967; Mintzberg, 1979; Moch & Seashore, 1981; Pfeffer, 1982;

Shapiro, 1977). In addition to these environmental constructs, the state of computing management for each organization (Kraemer, King, Dunkle, & Lane, 1989) could be investigated and treated as antecedents of information sharing. We propose a conceptual framework to address the expected relationships between a variety of these antecedent variables and the attainment of interorganization cooperation, as well as investigating the "second-level" effect of interorganization cooperation on projected consequences, or out-comes, of information sharing. These consequences are usually assessed as increased efficiency of organization tasks, increased effectiveness of output (Gillespie, 1991), and information system and partner-organization satisfaction (Ives & Olsen, 1983).

The antecedents shown in Figure 7.2 were chosen for several reasons. First, we felt it was necessary to identify two types of antecedents of information sharing and interorganization cooperation: those that are to some degree under the control of the project team members, as well as larger, organization-level constructs. Ease of communication, accessibility, and rules and procedures are proxies for organization variables that significantly influence associations among individuals and encompass many implicit and explicit aspects of an organization's strategy, structure, and culture (Peters, 1990; Pinto,

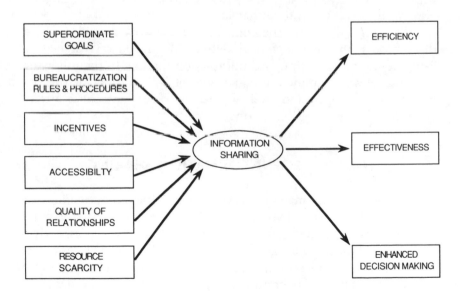

FIGURE 7.2. Antecedents and consequences of information sharing.

Pinto, & Prescott, 1993). For example, organizations that (1) permit their GIS departments to associate with other parties at their locations, (2) modify their work schedules to the demands of the project, and (3) develop their own rules and procedures to facilitate cooperation represent a different type of culture than those organizations that do not permit the same degree of latitude. Further, the acceptance of superordinate goals by team members should transcend and mitigate the role of individual factors. Additionally, there has been little organization-related field research utilizing superordinate goals as a method of facilitating cooperation. The quality of exchange relationships argues that positive interorganization relationships are marked by high levels of trust and minimal opportunism. Finally, research on resource munificence argues that organizations that operate under conditions of increasing resource scarcity are more likely to regard internally generated information as a source of power to be held over other parties. As a result, according to Pfeffer and Salancik (1978), there is a strong tendency to hoard, or refuse to share that information. The establishment of means for sharing information among organizations is a nonroutine task, whereas the goal is that actual sharing should become routine. Our choice of antecedents represents the need to balance past research findings that have dealt primarily with teams within an organization unit, on the one hand, with our objective of studying cross-functional teams working on nonroutine projects, on the other.

Figure 7.2 illustrates the conceptual framework that identifies a three-stage path analytic model delineating the factors argued to facilitate interorganization GIS information sharing. In the following sections, we will examine the importance of each of these factors, the importance of cooperation itself, and perceived outcomes, or results, of enhanced GIS information sharing.

THE NEED FOR INFORMATION SHARING

The responsibility for many project-based activities often overlaps, or is held jointly, between two or more GIS-using organizations. It has long been acknowledged, from an organization theory perspective, that these "difficult-to-assign activities give rise to such interdepartmental issues as cooperation, coordination, conflict, and struggles for power" (McCann & Galbraith, 1981, 60). To manage the development and implementation of these activities, a variety of integrating mechanisms have evolved, including task forces, liaison roles, and cross-organization teams (Galbraith & Nathanson, 1978; Lawrence & Lorsch, 1967; Mintzberg, 1979). These teams or task forces allow for lateral contact between multiple organizations (Dumaine, 1990). They tend to be temporary groups that exist for the duration of the designated activity. Because of a general unwillingness of some organizations to cooperate and share information willingly, the activities of these temporary task forces have become increasingly important for the long-term viability of an organization. Consequently, efforts must continually be made to develop policies and mechanisms that promote, rather than inhibit, cooperation across organization boundaries.

ANTECEDENTS OF INTERORGANIZATIONAL COOPERATION

Superordinate Goals

Classical organization theory originally established the importance of goals in organizations (Simon, 1964). Since that time much has been written on the concept of a goal for an organization (Kono, 1984), the purposes served by goals (Quinn, 1980), the multiplicity of goals in organizations (Raia, 1974), and the hierarchical nature of goals (Galbraith & Nathanson, 1978).

Every organization—and, indeed, every manager—has more than one goal that guides activities and actions. In theory, different organizations performing similar or complementary functions should possess complementary goals that are derived from a set of general, cross-organization objectives. In practice, however, overall goals are often broken down into specific objectives that conflict with, rather than complement each other. Consequently, in order for one organization to achieve its goals, another may be required to sacrifice, or at least compromise, its primary goals. Newman (1988) cautioned that a department's goal must be compatible with the activities of related departments so that it will not undermine the results of those activities or make them much more difficult. Consider, for example, the common conflict between state and local or state and federal agencies over GIS-data sets. Because the goals of these agencies are different and may often conflict, willingness to go out of one's way to share information collected by someone else is likely to be lacking. This situation results in a strong potential for duplication of data collection and maintenance efforts and may result in the implementation of systems that are each underutilized by their respective organizations. Members of each organization will argue the need for their own system and are often loath to cooperate or make available readily accessible information that may be needed by other agencies. Aware of such conflict, organizations and oversight groups are continually looking for ways to develop goals that can increase, rather than detract from, information sharing and interorganization cooperation.

One important, but often overlooked, type of goal in the study of cross-functional cooperation is a superordinate goal. As conceptualized in this chapter, superordinate goals refer to "goals that are urgent and compelling for all groups involved, but whose attainment require the resources and efforts of more than one group" (Sherif, 1962, 19). It is important to note, however, that superordinate goals are not a replacement for other goals that the various organizations may have; rather, they are an addition to existing organization goals. Through various controlled field and laboratory experiments, Sherif compiled impressive evidence indicating that when a series of superordinate goals are introduced into a conflict situation, intergroup conflict is reduced and cooperation is achieved. The essence of Sherif's theory is that competitive individual goals cause intergroup conflict but that superordinate goals give rise to intergroup cooperation that enhances group output. Specifically, when groups associate under conditions embodying shared goals or common purposes, they tend to cooperate as they work toward common goals. Because of this cooperation, system satisfaction is enhanced; thus, evidence suggests that a two-stage process is present. Superordinate goals are assumed to lead to enhanced data sharing which, in turn, is expected to result in an improvement in an organization's efficiency, effectiveness, and decision making.

Organization research on superordinate goals has tended to be conceptual in nature (Stern & Heskett, 1968) or has been conducted in experimental settings in which superordinate goals are manipulated (Johnson & Lewicki, 1969; Stern, Sternthal, & Craig, 1973). Therefore, there is strong evidence to suggest that superordinate goals can have a powerful effect on interorganization cooperation. What superordinate goals would cause numerous federal, state, and local government agencies and private sector organizations to work toward geographic-information sharing? Would such a superordinate goal be achieved by developing the concept of a nationwide library reference system for spatial data and by providing a distributed network infrastructure allowing ready transfer of large data sets? What lesser superordinate goals would enhance abilities as well as motivations to share?

Bureaucratization: Rules and Procedures

Rules and procedures refer to the degree to which activities or tasks on a project team are mandated or controlled. According to Galbraith and Nathanson (1978), rules and procedures are central to any discussion of interorganization cooperation because they offer a mechanism for integrating or coordinating activities, particularly those activities that cut across agency or department boundaries. Early organization theorists such as Taylor (1911), Fayol (1929), and Mooney (1947) relied on rules and procedures to link together the activities of organization members. They posited that coordination could be accomplished by simply establishing rules and procedures throughout the management hierarchy. Subsequent researchers (e.g., Gouldner, 1954; March & Simon, 1958; McCann & Galbraith, 1981) also discussed rules and procedures as a technique for coordinating activities, controlling behaviors, and maintaining the structure of an organization. At the department level, McCann and Galbraith (1981) described rules and procedures as the most common structural variable for assigning duties, evaluating performance, and minimizing the occurrence of conflicts between departments. Providing empirical support for this assertion, Reukert and Walker (1987) found that written or formalized rules and procedures have a significant positive relation to the perceived effectiveness of interdepartmental relations.

Galbraith and Nathanson (1978) noted that rules and procedures can be an effective method for achieving coordination between organizations. Specifically, they argued that standardization through rules and procedures is a useful integrating mechanism only when organizations operate under conditions of relative certainty and routine tasks. As an organization's design becomes increasingly complex, however, the effectiveness of rules as a coordinating device among departments decreases.

It is important to distinguish between the *concept* of bureaucratic control and the *effects* of bureaucracy on information exchange. This alternative assessment of bureaucratization argues that as bureaucracies evolve they tend to become stifling and territorial and will actually inhibit the flow of information across organization borders (Obermeyer, 1990a). The argument points to an organization's bureaucracy as one of the principal culprits in preventing free information exchange. This position is supported in the management literature by research within a marketing context that has found bureaucratic structuring to be damaging to exchange relationships and to exacerbate opportunism between organizations (John, 1984). While these arguments have merit, they do not negate but, rather, serve as a complement to our analysis of bureaucratization. We are here examining the concept of bureaucratization from the perspective of control mechanisms (rules and procedures) rather than directly addressing an organization's bureaucracy and its potential for noncooperativeness. In other words, when one agency sets up a series of procedural steps to ensure that their personnel will cooperate with other agencies, they are using a bureaucratic form of control. This is not to gainsay the potential negative effects that a large bureaucracy can have on cooperation between organizations but, rather, to argue that bureaucratization may also be viewed from another, more positive perspective.

A second conclusion relevant to our model is that rules and procedures have a means-end interrelation. While rules and procedures are developed ultimately to facilitate the accomplishment of desired ends, they also provide a means to establish cooperation among the individuals or departments charged with a particular task. The degree of rules and procedures are related to the degree of formalization in an organization. Recent studies by Deshpande and Zaltman (1987) and John and Martin (1984) found that increased formalization had a positive effect on the flow of information. Building on this research, Moenart and Souder (1990) suggested that increased formalization between departments produces a more harmonious climate. Therefore, it appears that rules and procedures should have a direct influence on the development of information sharing and cooperation between organizations. Thus, if the role of rules as the means does not facilitate cooperation, then the end state will most probably suffer.

Incentives

Another logical facilitator of an organization's willingness to share information with another has to do with perceived incentives. This argument captures the old question "What's in it for me?" that individuals and organizations frequently ask before engaging in any type of personal or professional commitment. Incentives suggest that an organization or its key members must perceive a payoff arising from the act of cooperating in an information-exchange relationship. Such a payoff may be in the form of creating a future bond of obligation or gaining some form of strategic or monetary advantage over rival organizations or agencies.

Under the incentive system, the willingness of one organization to participate in an information exchange lies in direct proportion to the other organization's providing the first organization with some scarce or necessary resource that it does not possess (e.g., money, access to important information, and so forth). When that other organization communicates its willingness to develop an economic-exchange relationship, the first agency must determine if the incentives are valuable enough to warrant the exchange of information for that resource. If the answer is yes, the information exchange will occur.

Accessibility

Accessibility usually determines the type and frequency of associations that occur between organizations. In our model, accessibility is defined as an individual's perception of his or her liberty, or ability, to approach or communicate with another individual from a different organization. Factors that influence the type and amount of association that occurs between organization members include an individual's schedule, position in an organization, and out-of-office commitments. These factors often affect the "accessibility" among organization members. For example, consider a setting in which an individual from a local government is physically located near an individual from a state agency. While these individuals are in close proximity to one another, they may rarely associate because of different work schedules, varied duties and priorities, and commitment to their own agendas. Within a manufacturing organization, Souder (1981) demonstrated that the lack of communication, lack of appreciation, and distrust that often exist between marketing and research and development units is fostered by normal time pressures, work deadlines, and some imbalance of power and prestige. These factors often lead to a perception of "inaccessibility" among the individuals involved. Building on a set of studies, Zaltman and Moorman (1989) found that associations between organization members enhance trust. Associations are easier when parties are accessible. Zaltman and Moorman (1989) suggested a causal link between associating and trust. Communication strategies and technologies that can help overcome lack of ability to approach another caused by differences in schedules, position in the organization, and out-of-office commitments should help build trust among people and organizations with the potential for sharing geographic information. In our model, previous research suggests that accessibility should directly affect the degree of cross-functional cooperation.

Quality of Exchange Relationships

Research supports the view that the quality of exchange relationships with potential exchange partners represents a significant criterion for evaluating cooperative information flow (Hunt & Nevin, 1974). The arguments underlying this issue suggest that member expectations of cooperation and information exchange are crucial elements in maintaining quality channel relationships. If these elements are characteristics of the system, different GIS-using organizations would be expected to fulfill obligations and would expect each other to desire coordination; high levels of satisfaction and morale should follow.

Following research by Dwyer and Oh (1987) that was originally compiled to address relationships among member organizations within marketing channels, we will analyze the concept of exchange relationships as consisting of three key dimensions: *satisfaction, trust,* and minimal *opportunism.* Satisfaction with an exchange partner is a significant issue when evaluating information-exchange relationships. Member satisfaction includes all characteristics of the relationship that the focal organization finds "rewarding, profitable, instrumental . . . or frustrating, problematic, inhibiting" (Reukert & Churchill, 1984, 227). Likewise, trust refers to a party's expectations that another desires coordination, will fulfill obligations, and will pull its weight in the relationship (Anderson & Narus, 1986). When one organization has built up a level of comfort and trust with another party, it is willing to engage in a more open exchange and sharing of relevant information (Deutsch, 1958). This argument suggests that the past history of working relationships will have an effect on future levels of trust and willingness to cooperate. Finally, opportunism is exemplified by distortion of information, failure to fulfill promises, and unwillingness to share what may be considered proprietary information. In effect, as one researcher suggested, opportunism is "self-interest seeking with guile" (Williamson, 1975, 26). When the condition of opportunism is prevalent between two agencies or organizations, there is little likelihood that needed information will be shared between the parties, as each or both of the exchange members view information as a base of power to be held over the other.

Resource Scarcity

The concept of resource scarcity refers to the idea that organizations must contend with a limited pool of resources when conducting their activities. Under certain circumstances, the resource level that is available to an organization may not be constraining but may allow a wide range of activities and options. This state is referred to as munificence. More often, however, organizations are constrained by limited budgets and the availability of technology, trained personnel, and other needed resources.

Research has long found that organizations and agencies operating under conditions of resource scarcity often tend toward the desire to maintain some form of control over other companion agencies. Indeed, the resource dependence model of an organization's power argues that a method by which one organization can exert control over another is through creating and stockpiling some resource that is scarce and is needed by another organization (Pfeffer, 1982; Pfeffer & Salancik, 1978). In the case of the current discussion, the needed resource may be information of the sort provided by GIS technologies. Because information is viewed as a form of power, the agency that possesses needed information is less likely to share it with another party for fear of losing some base of power in the exchange relationship.

On the other hand, when conditions of relative munificence operate, there is less emphasis placed on hoarding resources, either material or informational. Thus, the environmental condition of munificence offers an increased likelihood of information exchange, while perceived resource scarcity is likely to create the opposite effect, in which parties possessing information are not as inclined to make it readily available to other organizations.

INFORMATION EXCHANGE

Many concepts exist that encapsulate the meaning of cooperation between organizations. Among the various terms used to describe the notion of individuals working together to accomplish a specific task are (1) "coordination" (Argote, 1982; Van De Ven, Delbecq, &

Koenig, 1976); (2) "collaboration" (Trist, 1977); (3) "cooperation" (Schermerhorn, 1975; Sherif & Sherif, 1969); and (4) "integration" (Gupta, Raj, & Wilemon, 1986; Lawrence & Lorsch, 1967). The variability in terminology raises a critical question regarding what, if anything, the underlying concepts have in common. While each of the terms has a separate and distinct name, each refers to a similar and overlapping idea as evidenced by the commonalities in the definitions. The lowest common denominator that integrates the four concepts is *joint behavior toward some goal of common interest.* For purposes of this research, organization information sharing and cooperation are conceptualized as the degree, extent, and nature of interpersonal relationships among members from multiple organizations.

The need for information sharing and cooperation stems from the complex interdependencies among members of different organizations charged with complementary objectives. As Thompson (1967) suggested, greater interdependence requires a greater cooperation effort. Unfortunately, problems associated with cooperation between organizations result not only from the interdependence of work process and technology but also from conflicts over authority and jurisdiction among team members representing different departmental units. Thus, information sharing within an organization is essential in the implementation of decisions. It has been shown to promote productivity by helping individuals perform more effectively (Laughlin, 1978). Laughlin argued that organizations that are cooperating tend to (1) understand and be influenced by each others' interests and ideas, (2) seek and give information, (3) communicate about tasks, (4) more readily assist each other, and (5) rely on division of labor.

The previous discussion has established a link between antecedents and cooperation between organizations. Research has also shown that the antecedents may not have a strong direct link to project performance. Further, GIS information exchange appears to be directly related to the outcome measures of increased efficiency and client satisfaction. Thus, our model cooperation between organizations mediates between the antecedents and outcomes measures.

CONSEQUENCES OF CROSS-FUNCTIONAL COOPERATION

As Figure 7.2 indicates, greater sharing by organizations of geographic information is not simply an end to be sought for its own sake. Rather, we have posited a three-stage flow process in which information sharing represents the mediating link between the antecedent variables and outcomes in organizations. In effect, our model suggests that enhanced information sharing will lead to several desired outcomes that various organizations seek: efficiency, effectiveness, and improved decision-making ability. In addressing each of these outcomes, we will be making use of previous research in the area of the measurement of GIS use and effect developed by Gillespie (1991) and Zwart (1991).

One expected outcome of enhanced information sharing between organizations is greater internal efficiency of operations for each agency. When one organization is able to make direct contact with another party that possesses needed information, there is far less likelihood of replication of effort in creating databases. An organization's efficiency is enhanced through this sharing process. Increased efficiency from information sharing is measured, according to Gillespie (1991), by taking the difference in variable costs between the alternative methods for producing the desired output. In the case of geographic information, an estimate of the resources (labor, time, and money) needed to develop required GIS information in-house would be compared to the cost of retrieving such information from external sources with whom the organization has developed a cooperative relationship. If the difference is positive, then it could be argued that information sharing has resulted in increased efficiency.

A second outcome of information sharing is that of increased effectiveness in an organization. Effectiveness has been defined by Gillespie as the case where GISs "increase the quality of the output or produces a new output" (A-85). Gillespie has further suggested a three-step process for measuring effectiveness benefits. First, it is necessary to determine how the GIS output is different. In other words, what is being used or produced now that was not operational or available before? Second is the need to determine what effect each of the changes in GIS output has on the users of that output. One way to assess the changes that can take place is to examine the activities of organization members. Given the availability of new sources of information through cooperation between organizations, do members make use of this available information? If the answer is no, then it is likely that sharing has not increased effectiveness. Finally, the third step in assessing effectiveness is to determine the value of each of the effects on the users. Obviously, this step is the most complex and difficult because it requires that some figure of value be assignable to the set of effects.

The final source of outcome to assess is the effect on decision making of the new cooperative arrangement between organizations. Zwart (1991) argued that unless we can determine that utilization of geographic information has led to enhanced or better decision-making capabilities, its effect is minimized. Lucas and Nielsen (1980) reinforced the importance of improved decision making by suggesting that we need to assess effect by examining the degree to which information is utilized. They argued that for information to be fully utilized by an organization, it must not simply be referred to in decision making, but it must actually lead to changes in an organization's values or in managerial decisions. As a result, if information that is shared between organizations does not lead one party to reassess decision priorities or value structures actively, the third criterion of outcome (decision making) has not been fully addressed.

CONCLUSIONS

In this chapter we have sought to create a context for providing a greater understanding of the problems associated with interagency information sharing and of some key factors in helping to mitigate these difficulties. Further, by way of illustration, we have developed a conceptual model for the investigation of managerial and organization-level constructs that can serve as facilitators of the sharing of geographic information between organizations. As Figure 7.2 demonstrates, the underlying framework upon which we have based our set of research propositions involves a series of hypothesized causal links between antecedents of information sharing and its projected outcomes for organizations. From the model, it is expected that enhanced information sharing will result in a number of positive outcomes for an organization. These benefits include greater efficiency, effectiveness, and improved managerial decision making.

The study of information sharing within the context of GIS environments is still in its infancy. By making use of current organization theory and marketing channel research, we have attempted to create a framework for a better understanding of some of the causes of cooperation between organizations sharing geographic information. These streams of research have allowed us to make some supportable propositions concerning the set of causal flows between antecedents of information sharing and the likely results for an organization engaged in a cooperative venture. It is hoped that as a result of a better understanding of these linkages, managers and members of organizations tasked with the responsibility of developing cooperative relationships will be in a position to facilitate geographic information sharing to the advantage of all concerned organizations.

❖

GISs and the Strategic Planning Process

The purpose of this chapter is to demonstrate the integration of geographic information into an organization's overall strategic planning process. We will show that a GIS can be used as an important and central element in the development of strategic objectives and long-range plans. The concept of strategic planning will be defined and we will propose a general model of strategic planning that will serve as the basis for gaining a better understanding of all relevant elements in the strategic planning process. Finally, we will analyze in some detail the role of GIS in developing comprehensive strategic plans, suggesting that the type of information provided by a GIS makes it uniquely capable of enhancing the planning process for a variety of public and private organizations.

Because of the increasingly complex nature of the environment within which organizations operate, a variety of new and challenging demands face today's managers. Novel and more complicated problems have given rise to a better-educated work force using new technologies to help them manage their professional responsibilities. The introduction of computer technologies and other systems-integration tools has helped to increase the speed of communications, thereby quickening the pace of day-to-day activities. Similarly, the world outside the organization has become more complex and faster paced. Issues that might once have been viewed from a narrow or parochial perspective are now routinely address as part of a global view.

As an organization's environment—both internal and external—has become more complex, an organization's decision making has become more challenging. Strategic planning (sometimes referred to as strategic management) has developed as a management tool over the 1970s and 1980s to meet the challenges posed by the increasingly complex environment of an organization. For a number of public and private organizations, a GIS is a uniquely qualified tool to aid in long-range planning and objective setting. For these organizations, geographic information is a vital element in their planning process. Whether the organization is a local government concerned with new residential development and subsequent questions of rezoning and infrastructure expansion or is a paper company intent on ensuring a supply of timber for its long-term operations, geographic information of the type provided by a GIS can serve as a key element in setting a strategic focus for these organizations.

Because GIS is often so central to the process of strategic planning, in this chapter we will focus on the relation between an organization's planning process and the role that a GIS can play in these activities. We will develop, in some detail, the basic elements in a strategic plan. Further, we will address a generic model of strategic planning to show the interrelatedness among the various organization and human factors that can influence the planning process. Finally, we will discuss the specific stages in the planning process and the reasons why geographic information of the type provided by a GIS can be so beneficial.

WHAT IS STRATEGIC PLANNING?

One of the earliest comprehensive and relevant definitions of planning was offered by Scott (1963), who stated that "planning is an analytical process which involves the assessment of the future, the determination of desired objectives in the context of that future, the development of alternative courses of action to achieve such objectives and the selection of a course (or courses) of action from among those alternatives." Scott's definition is important because it identifies several of the key planning elements and activities. These are important points that need to be underscored in his definition because they cut to the heart of any approach to strategic planning. First, planning is a *process*—that is, it represents a dynamic, ongoing effort on the part of the organization and its members. Planning is not a static event. It does not simply occur at well-defined intervals, is engaged in for a prescribed time period, and is then abandoned. Effective strategic planning is a robust, continuous activity on the part of the organization as it acknowledges that change is unpredictable, continual, and potentially highly significant to the organization's future operations. Further, because it is analytical, the planning process is essentially rational in its intent and its efforts.

The second important element in the definition is the idea that planning is *future directed*, with the intent focused on determining the appropriate objectives for an organization within the context of possible significant changes in the organization's operating environment. Finally, strategic planning is *flexible*. It requires planners to anticipate a variety of possible scenarios, develop alternative courses of action to deal with these scenarios, and—depending upon the environmental threats and opportunities encountered—select the appropriate course of action that can most effectively address the organization's concerns.

The major innovation of strategic planning is its underlying assumption that the world is uncertain and unpredictable. An organization must understand its uncertain environment (both internal and external) if it is to be successful in accomplishing its mission. Understanding the environment is an ongoing task, since the environment is subject to change; hence, the genesis of the phrase *strategic management*. Figure 8.1 illustrates the various components of the strategic planning process. This figure is important because it illustrates the interrelatedness of the component elements and characteristics of the planning process. These elements include (1) the organization's *analysis process*, (2) its *expected future*, (3) its *perceived competences*, (4) *top manager's preferences and values*, and (5) *stakeholders' priorities and power*, all of which affect (6) the organization's *strategic plan*.

The Analysis Process

The *analysis process* refers to the formal methods by which the organization engages in strategic planning—that is, the information that staff members deem relevant to their operations and that they choose to collect, the specific steps that they take in developing their strategies, and the analytical process based on hard data (e.g.,

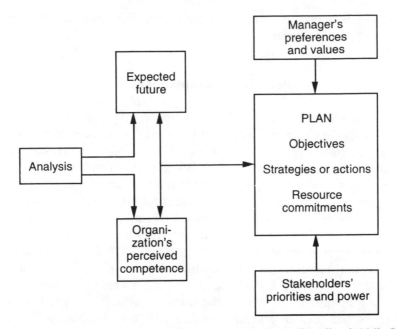

FIGURE 8.1. Steps in a generic planning process. From Camillus (1986). Copyright 1986 by Lexington Books. Reprinted by permission.

computer-provided analyses) or generally consisting of "soft" data based on word-of-mouth or other qualitative information.

The Expected Future

The *expected future* of an organization simply means that the organization develops a variety of scenarios to assess as many possible future states as can be envisioned by the planners. Essentially, the planning staff are asking a series of "What if?" questions in order to cover every possible future contingency. As an example, one of the contributing events that led many companies to develop their own strategic planning departments was the series of OPEC-generated "oil shocks" of the 1970s. So many companies were seriously hurt by these embargoes and price hikes that they determined never to be caught unaware in the future. Although the *expected future* means just that (the future that seems most probable), it is also important to develop several alternative scenarios and analyze appropriate responses of organizations in order to reasonably ensure the likelihood of not being surprised by future events.

Perceived Competences

In tandem with the expected future, strategic planning needs to acknowledge an organization's set of *perceived competences*. What is it that a particular organization does well? If the firm is private, the question that needs to be asked is What is this firm's competitive advantage over other companies? What has the firm done, and what does it continue to do, that makes it successful relative to its competition? If the organization is public, such as an agency of a local or county government, it still must examine its activities to determine in what areas it possesses a set of distinctive competences. For example, some towns pride themselves on the recreational elements of their community (water sports, skiing, golf, etc.). Other towns may take justifiable pride in their low-crime and public safety records. The logic behind asking, "What does an organization do well?" is to establish a framework for planning. Sound strategic plans almost always are intended to build upon current strengths rather than to reinforce inadequacies.

Top Managers' Preferences and Values

Obviously, a serious contributing factor to the focus of a strategic plan will involve the preferences and vision of the organization's chief administrators. Top management has the ultimate responsibility for the welfare of the organization and, because of its power, can have a tremendous effect on the future direction of the organization. In fact, one of the long-acknowledged duties of senior administration is to set long-term goals and strategies for organizations. As a result, it is easy to understand how the values and beliefs of top management will have a great influence on the types of plans and strategic directions taken by an organization.

Stakeholders' Priorities and Power

A *stakeholder* is defined as any group or individual with a potential interest in or effect on an organization. In other words, anyone who has a "stake" in what the organization proposes to do is referred to as a stakeholder. Stakeholders may have a great deal of power in some organizations (e.g., a large stockholder of a publicly held company), while being relatively powerless in others. One question that it is necessary to ask prior to committing to a strategic direction and developing subsequent long-term plans is the nature of the organization's stakeholders. Are they powerful? Must they be consulted before strategic planning takes place? To what degree are they capable of altering strategic plans and, in effect, redefining the organization's mission?

An organization must assess the influence of its major stakeholders on a continual basis to ensure that they (the stakeholders) are kept apprised of the directions that the organization takes. Within public organizations, many towns hold town-meetings or regular meetings of their city councils as a forum for stakeholder groups to present their ideas, to register their support for or objections to strategic choices, and to attempt to influence public policy. For the public sector, stakeholder groups may take on many forms, including small, localized special interest groups as well as large nationally based efforts (e.g., MADD, NRA, etc.). The groups work to influence public policy at all levels and are, to varying degrees, often successful in setting strategic agendas and influencing long-term planning.

An Organization's Strategic Plan

The final element in the generic planning process is the actual development of the strategic plan, which is subject, as we have noted, to the myriad groups and individuals who are to varying degrees capable of influence over these plans. As you can see from Figure 8.1, there are three basic elements to the creation of a strategic plan. First, the organization develops a set of objectives or goals for the future. These goals may include short-term goals (milestones to occur in less than 1 year), medium-term goals (from 1 to 3 years), and long-term goals (from 5 years to well over 20 years in the future). The objectives signal the planning team's analysis of the likeliest future directions for the organization, based on their interpretation of current data and trends. Following the establishment of these objectives, a series of specific strategies or steps are agreed to. These strategies consist of the activities that the organization believes are necessary to achieve long-term goals. For example, if one goal for a small, seaside town is to promote additional tourism and development over the next 10 years, one intermediate strategy necessary for realizing this objective might be to rezone oceanfront property for multiple dwellings and begin infrastructure expansion and dockside renovation.

Finally, the third important step in the strategic plan is an organization's commitment of resources to aid materially in supporting the strategies necessary to achieve long-term goals. Programs and strategies that are not materially supported by top management are almost invariably doomed to failure. Providing resources in support of strategic choices is often a difficult prospect, particularly in times of tight budgets and shrinking revenue. Yet, difficult times are also often the test of an organization's commitment to its strategic objectives. When difficulties set in, the easiest avenue to follow is to revert back to old, familiar patterns, including outmoded plans and strategies. This approach, while often convenient in the short-run, is almost always disastrous for the long-term viability of an organization. For example, in the late 1980s and early 1990s Congress has passed three budget balancing acts (including the well-known Gramm–Rudman act) and, when financial difficulties have threatened, has violated or overturned each piece of legislation. While often alluring, the short-term "quick fix" is usually a serious long-term disability for an organization.

Based on the interaction of each of the elements in strategic planning, we begin to see the emergence of a complex picture in which a variety of factors can influence the development of an organization's objectives and strategies. The equation is made even more complex by the existence of a number of stakeholder groups, each of which has a strong vested interest in influencing the development of an emergent strategy. The model shown in Figure 8.1 offers some indications of the complexity of the true strategic planning process. However, while complex, this model does not offer a comprehensive outline of the actual steps that managers need to take in order to develop a viable strategic plan. The following sections of this chapter will discuss the main elements of a strategic plan—in particular, demonstrating the role of geographic information in creating long-term goals and the strategies to realize those goals.

THE PROCESS OF STRATEGIC PLANNING

While the potential payoffs from effective planning can be tremendous, the strategic planning process itself poses a series of important challenges for managers. As Camillus (1986) has noted, the "design of the process poses a particularly significant challenge in that it is an important means by which rational, economic analyses can be integrated effectively with behavioral, political realities" (59). This statement underscores a key aspect of the planning process: that it does not consist entirely of the use of analytic methods to arrive at the best (profit-maximizing) alternatives and objectives. Rather, strategic planning recognizes the inherent interaction of economics with political and human realities. Consequently, in engaging in the strategic planning process, three important questions must be addressed prior to active information search and objective setting.

1. *What* should be analyzed? This is the rational/economic perspective that seeks to determine what data are important for developing a strategic plan.

2. *Who* should conduct the analysis? The political realities of most organizations suggest that certain individuals have a greater stake in long-term objectives—and, consequently, their decisions will carry greater weight in strategy development and goal setting.

3. *When* and how often should analyses be conducted? The timing perspective of strategic planning is also crucial. What should be the strategic planning cycle for our particular organization?

Bases for Strategic Analysis

The typical strategic planning process involves an in-depth scan of both internal and external elements of an organization. There are three generally accepted dimensions that must be addressed through this scanning process: *the industry situation, the competitive situation,* and an internal analysis involving *the organization's own situation* (Thompson & Strickland, 1987). Each of these dimensions will be addressed in turn. It is important to point out up front, however, that the strategic situation faced by private and public organizations is clearly different relative to industrial and competitive forces. Certainly for private (profit-seeking) firms, an in-depth market and competitive analysis is essential for long-term survival. On the other hand, municipalities and other government agencies are not, by their very nature, threatened with the same type of competition and concern for profit maximization. As a result, readers must carefully consider each element in the strategic scanning process to determine its particular applicability to their own organization's situation. Some of these elements may simply not be applicable to all types of organizations. Nevertheless, this framework offers an important starting point for strategic planning by emphasizing some of the key aspects of an organization's environment, both internal and external. Further, it will be shown that geographic information of the type generated by many GISs can be central to developing a better overall understanding of the environment within which an organization is operating.

Industry Situation Analysis

The industry situation analysis involves an examination of overall industry structure, direction, economics, and its long-term viability (attractiveness). Obviously, for public agencies and municipalities, issues of economics and attractiveness take on a different meaning than what they have for private firms. Nevertheless, the basic elements of a situation analysis are equally important and have meaning within the public sector. The term "industry" can be broadly defined as the set of organizations that are so similar that they are drawn into competition with each other for the same client base. The underlying purpose of the industry situation analysis is to gain a sense of the long-term potential for effectively competing within an industry. As a result, the industry situation analysis requires planners to take a clear look at their own organization relative to expected and potential changes within the overall industry or economy. Based on the identified characteristics of an organization and its external environment, is it more or less likely to remain a strong competitor into the future?

Table 8.1 gives some examples of the types of questions that planners need to ask in conducting an industry situation analysis. That is to say, in asking how the industry is structured, planners attempt to determine whether the industry is relatively closed or if it can be readily entered. For example, the forestry and paper products industry may be thought of as relatively closed in that it has high barriers to entry. To become a viable competitor, a company must not only engage in a capital outlay for plant and equipment (sawmills, transportation, etc.), but must also invest heavily in land purchases to ensure a steady supply of the timber that serves as the principal raw material. Because of these high barriers to entry, a central question that must be resolved is the degree of attractiveness that the industry still holds in spite of high start-up costs.

TABLE 8.1. Important Questions in the Strategic Planning Process

Industry situation analysis

How is the industry structured? Are barriers to entry high or low?
What general trends or driving forces can be discerned?
What are the key success factors in this industry?
What long-term strategic issues does this industry face?
Should we be in this industry? Why or why not?

The competitive situation

What competitive forces exist? How strong are they?
What do we know about key rivals? Their strategies and competitive strengths?
Where will future competition come from? What will it be like?

The firm's own situation

How well is our present strategy working? Will it require minor adjustments or
    wholesale changes?
What does our SWOT analysis tell us? What are we doing right and more
    importantly, what are we doing wrong and how can we fix it?
What is our competitive advantage?
What specific strategic issues do we need to address?

Some examples of other things that must be considered in performing an industry analysis include an identification of key success factors in the industry, the long-term trends the industry faces, and a general question of whether the firm should even continue to operate in the industry. Again, to use the example of the forestry company, one potential long-term trend that could bode ill for firms of this type is the enhanced concern for natural resources by vocal advocacy groups such as the Audubon Society and Green Peace. The power of these groups poses a serious threat to the long-term profitability of natural resource-based companies. To illustrate, the continued exploitation of the northwest timberlands has been severely restricted as a result of the threat to the Spotted Owl. The federal government has at present made the determination that preserving this endangered species outweighs the potential loss of thousands of forestry jobs.

Public organizations are also in a position to conduct their own version of an industry situation analysis. While they do not generally face the same types of pressures toward profitability, they are just as concerned about the long-term survival of their agencies or municipalities. Consider, for example, the case of the March of Dimes, which was established originally as a charity to help with childhood polio. In the wake of Dr. Salk's polio vaccine, which essentially eliminated the threat of polio, the March of Dimes was forced to perform an industry situation analysis to determine where its efforts would now be needed. As a result of this analysis, the March of Dimes chose to involve itself with birth defects. Public organizations and governments need to engage in industry scanning and analysis for their long-term survival to the same degree that private firms do. The difference, as illustrated in Table 8.1, lies in the fact that the type of questions that public organizations need to concern themselves with are often markedly different from those of private firms.

The Competitive Situation

While the industry situation analysis forms a backdrop for in-depth examination of an industry's competitive posture—sources of competition, barriers to entry, key industry success factors—the competitive analysis helps organizations to engage in a more focused search for advantage vis-á-vis their direct competition. In analyzing the competitive situation, three lines of questioning are key:

1. What is the nature and strength of the various competitive forces?
2. What are the competitive positions and strengths of key rivals? What are their strategies? What do they do better than we do?
3. What can we expect our rivals to do next?

The importance of the competitive analysis segment of the overall strategic planning process cannot be overestimated. It is by conducting an analysis of competitors that an organization comes to a better understanding of its own position in the industry. In other words, it is through an organization's efforts to gain an understanding of the strengths and weaknesses of its competition that it is able to gain a greater understanding of its own strengths and weaknesses. Even firms that are in attractive industries may find themselves unprofitable because they have allowed themselves to be placed in a weak competitive position against aggressive rivals. Hubris, poor planning, and willful blindness can all cause an organization to downplay competitive-position analysis that will almost inevitably result in a severely curtailed market. To illustrate: consider the real example of a city with a harness-racing track that has traditionally been the source of enhanced revenue for the municipality. Under a series of misguided assumptions that (1) casinos will never work, (2) casinos and race tracks are nonsubstitutable forms of entertainment, (3) the "real money" can only be made in racing, and so on, the city council chose to ignore reports of a neighboring community's efforts to construct legalized gambling casinos. The neighboring community's casinos became so popular that they severely lowered racing revenues to the point where the track had to be temporarily closed. While this is one example, in other cases cities have suffered a real loss of revenue through a competitive force that they chose not to take seriously until it was too late.

The above example highlights an important point that must be stressed about the benefits of competitive analysis—namely, its relevance to public administration organizations. The point can justifiably be made that a town planning board or city agency is not threatened by the actions of "rivals" in the same manner or to the same degree as private or not-for-profit organizations are. If public-administration GIS managers do not face the same sorts of competitive pressures, what is the benefit to them in conducting competitive-situation analysis? In order to answer this question, it is important first to understand that "competition" arises from a number of sources and in relation to a number of functions—other than simple profit making, as engaged in by privately held firms. The above example of a town using harness racing as a form of municipal-revenue generation is a case in point. Should another municipality within a convenient distance engage in similar activities, their actions could pose a competitive threat to the town that has promoted racing. Likewise, one community's decision to raise property taxes to a higher level than those of other neighboring communities could lose considerable support if residents perceived no difference in municipal services offered, and, hence, relocated to areas with lower taxes. This is consistent with the Tiebout hypothesis, previously discussed in Chapter Six.

The point that must be emphasized is that communities and other types of public administration may not face the same pressures from competition that has become synonymous with American business. This does not mean, however, that they do not face their own forms of competition from neighboring communities as well as from other public agencies at state and federal levels. As a result, it is important that both private and public sector GIS managers consider the potential effect of competitors on their operations and, consequently, engage—to a varying degree of sophistication—in some form of competitive analysis.

Analyzing the Competitive Situation:
The Five-Forces Model

An extremely useful tool for gaining a better understanding of the nature of competitive pressures is through the five-forces model developed by Michael Porter of the Harvard Business School. Porter argues that any organization must be aware of five distinct forms of competitive pressure in the marketplace:

1. *The jockeying for position among rival organizations in search of competitive advantage.* Each firm attempts to employ its own brand of competitive strategy in order to gain a favorable position relative to its chief rivals. This "competitive edge" becomes the key to long-term success by maintaining an advantage over other rival organizations.

2. *The intrusions and threats from the substitute products of organizations in other industries.* Competitive forces can also arise from the threat of substitute products from other industries. For example, if soft-drink manufacturers price their products too high, they run the serious risk of losing market share to fruit juice or powdered drink producers.

3. *The potential for entry into the industry by new competitors.* If the barriers to entry into a market are perceived to be low, the industry's attractiveness could increase, encouraging new competitors to enter the market. Further, this threat of new entrants can have a constraining effect on firms already in the marketplace. The airline industry offers a good example of the need for constraining influences. Owing, in part, to perceived inequities in fare structures and monopolies within certain markets, the airline industry was deregulated in the late 1970s. Deregulation led to a large number of new firms entering the airline business in the late 1970s and early 1980s. In order to gain a competitive advantage, a series of fare wars began, which has had the effect of shaking out a number of small and less financially sound airlines.

4. *The power and bargaining leverage of suppliers.* Important suppliers of all types of raw materials for an organization can influence not only that firm's ability to conduct business but also the strategic posture adopted by the company. The makers of semiconductors have such an influence on the operations of calculator and computer manufacturers that at one time IBM purchased a significant share of Intel Corporation's stock (Intel is the world's largest manufacturer of semiconductor chips). It is also important to note that a supplier's

bargaining position and power over an organization often lies in direct proportion to the lack of substitutes for the supplier's product. For example, the power of suppliers of glass bottles over soft-drink manufacturers is constrained by the availability of the use of aluminum and plastic containers as substitutes.

5. *The power and bargaining leverage of customers.* Just as powerful suppliers can influence the operations of firms within an industry, so too can powerful customers. When customers are few in number and are actively being pursued by a number of rival firms, they have a great deal of power in negotiating advantageous deals. Further, there are examples of cases in which a firm has signed a series of exclusive deals with suppliers for their products. Because the one company now poses as an exclusive customer, it can shape the future direction of the company for which it is the chief customer.

Figure 8.2 illustrates Porter's five-forces model. One point to note is the tremendous degree of interaction that these forces often have on each other as well as on the organization. In conducting a competitor analysis, the organization must be aware of the nature of each of these threats and the related fact that energy spent attending to one threat may actually encourage another threat to appear. As an example, let us assume that XYZ Corporation perceives that its chief rival has a technological advantage over it through more sophisticated manufacturing procedures. In an effort to reduce that technological advantage, XYZ Corporation develops a new low-cost procedure that allows it to be competitive with this chief rival. The advantage is that the company has negated the technological edge in manufacturing that the rival held. However, because of the new low-cost breakthrough, the first company has, in effect, lowered a major barrier to entry into the market and to new competition and substitute products. As a result, owing to the interaction of the five forces of competitive pressure, the first company may now actually be worse off than it was before.

An Organization's Own Situation

In addition to conducting a detailed external analysis of both the nature of the marketplace within the industry and the competitive stance of the organization relative to its competitors, it is extremely important that the organization devote sufficient time to performing an internal analysis. Up to this point, the focus of the organization's

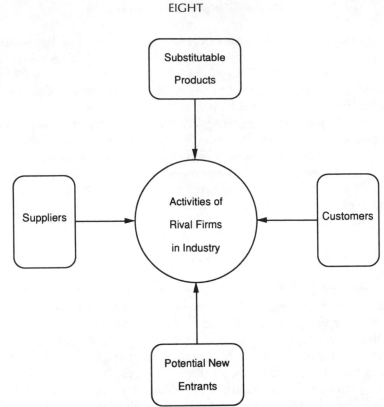

FIGURE 8.2. The five-forces model of competition. Copyright 1979 by the President and Fellows of Harvard College; all rights reserved. Adapted and reprinted by permission of *Harvard Business Review*, "How Competitive Forces Shape Strategy" by M. E. Porter, March/April 1979, 57(2).

analysis has been on the external arena, assessing the industry as a whole as well as the firm's competitive posture. At this stage, the spotlight shifts to the organization itself. Simply put, an internal-situation analysis requires the organization's planners and top management to conduct a detailed assessment of the current state of their organization and to address some compelling questions. How well is the present strategy working? What do we do well? In what areas do we need improvement?

The primary methodology that drives the internal analysis goes by the acronym SWOT and refers to the organization's assessment of its internal *strengths* and *weaknesses* as well as its external *opportunities* and *threats*. First and foremost a SWOT analysis consists of an honest appraisal of the current state of the organization and can be extremely useful for sizing up the company's activities and operations. The underlying concern is that the analysis be conducted for the purpose, not of validating the conduct of the organization, but of analyzing the companies strengths *and weaknesses* as candidly as possible. A SWOT analysis is not intended to be a self-congratulatory process, but is intended for reaffirmation. It is a highly useful method for staying abreast of an organization's activities.

Table 8.2 gives some examples of the types of issues that are important in performing a SWOT analysis. The firm's answer to these questions will go far toward pointing the organization in the direction of remedial activities to correct defects and solidify strengths. It is important that, in answering these questions, as wide a range of involved managers as possible take part in the process. The obvious advantage of using multiple respondents to assess an organization's current state (its strengths and weaknesses) is that some concerned parties or individuals, particularly from the top management group,

TABLE 8.2. Examples of Key Questions in the SWOT Analysis

Potential strengths?	*Potential opportunities?*
Distinctive competences?	New markets or niches to enter?
Sufficient Financial, human, and Raw Material Resources?	Diversification into new product lines?
Technological innovativeness?	Decline of competitor's position?
Manufacturing efficiency?	New customer expectations?
Good management?	Development of acquisition strate-
Cost advantages?	gies?
Industry leadership?	Vertical integration?
Good functional integration?	Advantageous demographic changes?
Comprehensive information system (including GIS)?	Potential for market or resource development
Potential weaknesses?	*Potential threats?*
No clear direction?	Likely entry of new competitors?
Falling behind the experience curve in research and development?	Changing customer needs or tastes?
Narrow product line?	Encroachment by substitute products?
Obsolete or inefficient production?	Enhanced bargaining position of suppliers?
High cost of operations?	Enhanced bargaining position of customers?
Poor worker/management relations?	Adverse demographic changes?
Inadequate functional skills?	Market growth slowing, stagnant, or declining

may be ignorant of or willfully blind to weaknesses within company operations.

Organizations have differed in their approach to conducting, and in their subsequent use of a SWOT analysis. Some firms have compiled detailed checklists similar to, but more comprehensive than, that shown in Table 8.2. Following the creation of these checklists, the company's managers perform their SWOT assessment using some form of scale (e.g., a 5-point scale where 1 = "needs improvement" and 5 = "outstanding"). The strategic planning team then conducts an analysis of how their own organization stacks up against the checklist and performs some "best guess" SWOT analyses on their chief rivals. The use of SWOT analysis in this manner has some advantages in that it allows for comparisons between an organization and its competition. The company can make some tentative determinations regarding the way in which they stack up in the industry. While this is one approach to SWOT analysis, the important point to remember is that no matter how the analysis is conducted, the benefits to be derived from actually engaging in the process can be extremely eye opening, particularly for firms that have never made this type of systematic assessment.

STRATEGIC PLANNING AND GIS

What, then, is the role of geographic information in the strategic planning process? The obvious response to this question is that for many organizations, the GIS has become intimately tied to the entire strategic planning process. In effect, it becomes an important element in this process. One of the points we have tried to make throughout this chapter is that the strategic management process is intensely information-dependent. In order to engage in sufficient industry and competitor analysis, a great deal of both internally and externally generated information is required. One of the generally accepted strengths of the GIS is its ability to collect and sort the myriad pieces of information, thus assisting planners tactically (concerning day-to-day decision making) as well as strategically. The strategic use of the GIS is seen every day in both public and private organizations. For example, urban-planning agencies—tasked with the responsibility of developing comprehensive and sound policies for zoning, urban renewal and development, land reclamation, and so forth—have found that the GIS offers an efficient and effective technology for providing the type of information so important in the decision-making process. The GIS has become a central storage facility for the thousands and thousands of diverse pieces of information that can be stored against the future possibility of their being needed in making strategic planning decisions.

The role of the GIS in strategic planning becomes even more apparent when we reexamine the various elements of the strategic planning process as set forth in this chapter. We have suggested that the three key questions first needed to be addressed are, *What* should be analyzed? *Who* should conduct the analysis? *When* and how often should the analysis be performed? We can see that the use of a GIS gives the planning group tremendous flexibility in addressing each of these crucial questions. First, if an organization has taken adequate time to ensure that it is collecting and storing all necessary information for its information system, the question of what should be analyzed becomes much easier to deal with than in other situations in which the organization does not have any form of comprehensive information system. In the latter case, the company may be forced to make decisions based on insufficient information. Because the company has not taken the steps to ensure an adequate information storage and retrieval system, the whole question of what should be analyzed becomes somewhat academic. Such an organization is forced to analyze whatever information it has available, the result of which is often making "seat of the pants" decisions based on past experience of gut instinct rather than on analysis of sufficient information.

Another benefit of the GIS refers to the question, *Who* should be conducting the strategic analysis? Most existing research points to the fact that political realities often place certain higher-level individuals in the position of exerting more influence over the decision process. While it is true that higher-level individuals can often influence the nature of goal setting and information search, the availability of a GIS has something of an egalitarian effect on the organization. In other words, through the creation and use of a GIS, information becomes more accessible to a wider range of people. Rather than making decisions in a "black box" manner, in which the majority of the organization is unaware of how decisions are made, a GIS puts the sources of information literally at an individual's fingertips. As a result, a wider range of people can make use of the information that is necessary to engage in effective decision-making. Further, a GIS can serve as a check on individuals misusing the power of their position. When the information on which a decision was based is instantly available to a number of people, there is a greater that these individuals will challenge basic assumptions of the decision maker.

Finally, a GIS also answers the questions of when and how often analyses should be performed because it offers strategic planners the capacity for continual planning. Because a major information source is always available and instantly accessible, the planning process has the capacity to operate continually. As new data are fed into the storage system, information is constantly upgraded. These upgrades are intended to maintain the effectiveness of a major planning tool. As a result, strategic planning has the capacity of being performed on a continual basis, the only limitation being the amount of new information fed into the system, the frequency with which the system's database is upgraded, and the availability of trained personnel within the organization to access this information and make use of it in the planning process.

As a final point about the role of the GIS in the strategic planning process, we can see that geographic information is useful for conducting both a situation analysis as well as an internal analysis of an organization. The industry analysis requires that an organization make a series of determinations about the nature of its external relationship within the industry. How will the organization compete? Is the industry still as attractive to the organization as it once was? Some forms of geographic information provide important answers to questions about industry status. For example, a hydro-electric facility may determine from census data and water flow analysis both that its client base is shrinking and that, because of recent engineering projects, water levels are falling to the point where it will become increasingly costly to continue generating electrical power. These determinations could signal that the industry is rapidly losing its attractiveness, which could thus lead to significant strategic reorientation.

Further, the GIS can provide important information as part of an organization's internal SWOT analysis. Through the analysis of geographic information it may be found that there are significant threats to a city's infrastructure from old systems or neglected maintenance. As in the case of Chicago's recent downtown flooding, information had been available to suggest that there was imminent danger of catastrophic flooding, which was generally disregarded until too late. New York, as well, is currently facing similar problems with its infrastructure (most notably, bridges). The GIS can provide not only important information for municipalities regarding future opportunities in the areas of development but also warnings of potential threats from a number of sources.

CONCLUSIONS

In this chapter we have examined some of the key elements in the strategic planning process for both public and private organizations and the central role that a GIS can play in this process. Strategic planning has become increasingly important for most modern organizations within both the public and private sectors. It gives these organizations a "sense-making" device that can help managers understand the relationship of their organization to the external environment (including competitors). Further, it offers these organizations the opportunity to maintain a proactive, rather than a reactive, relationship with their environment. There is an old saying, "Those who fail to plan, plan to fail." The central thrust of this saying for most organizations is that strategic planning is no longer a luxury but a necessary part of their activities. For most public and many private organizations, geographic information of the type provided by the GIS is a powerful tool in the planning process. This chapter has attempted to explicitly link the process of strategic planning with the functions of the GIS, demonstrating the essential and growing interrelatedness of their operations.

NINE

❖

Ensuring the Qualifications of GIS Professionals

The rapid proliferation of GIS adopters has raised concerns within the GIS community about ensuring the competency of GIS professionals. Certification is the strategy most commonly suggested as a means of achieving this objective. It involves the specialized training of individuals who are ultimately tested on what they have been taught. Those who pass the examination are certified. Accreditation refers to a profession's recognition or endorsement of certain curricula or training programs as meeting the educational criteria required to enter and practice the profession. Both of these strategies—certification and accreditation—have advantages and disadvantages. In this chapter we will examine certification and accreditation as alternatives to the status quo and will raise questions that the GIS community as a whole must consider before making any decision. We recommend first addressing this issue by asking the question, How can the GIS community best ensure the competency of its professionals?—rather than beginning with the assumption that certification is the only means to this end. Since the primary objective of this chapter is to stimulate discussion on how the GIS community can best ensure the competency of its professionals, we will deliberately avoid making any recommendations about which strategy the GIS community should follow to ensure the competency of its professionals, suggesting instead that the GIS community formally consider this topic in order to achieve a resolution. We will conclude the chapter with a discussion of practical approaches organizations can adopt to be sure of the qualifications of the GIS professionals they hire.

Geographic information systems are growing in demand and are proliferating among both public and private organizations. One of the outgrowths of this increased demand is a shortage of GIS profession-als—a fact that has frequently been noted in the GIS literature. Members of the GIS community have expressed concern that the shortage of GIS professionals will lead to a situation in which poorly trained individuals may assume positions for which they are unquali-fied. The employment of inadequately trained people in the GIS field has the potential to cause the failure of GIS projects, which, in turn, has the potential to damage the credibility of GISs and the profession itself.

As a result, there is within the GIS community a growing concern about finding a mechanism to ensure the competency of individuals advertising themselves as GIS professionals. The most commonly suggested strategy to achieve this goal is the development of a certifica-tion program for individuals who wish to hang out their GIS shingle. In fact, the surveying profession, which employs certification to assure competency among its members, has made inroads in Georgia and California toward the achievement of this goal. However, certification is not the only mechanism designed to guarantee the competency of professionals. Accreditation of programs of instruction in a field is also a well-accepted means to this end.

As a practical matter, the question of competency warrants serious discussion, particularly for organizations implementing GISs. In some instances, organizations will hire GIS professionals from outside the company or agency. In other instances, the organization will train personnel in-house. Either way, competency is a serious concern.

In this chapter we will raise questions regarding both certification and accreditation as alternatives to the status quo. We will begin with a brief discussion of the current situation in the GIS community, then discuss two alternatives—certification of individuals and accreditation of programs—as a means of ensuring the competency of GIS profession-als. The discussion of certification includes experiences from two other professions—medicine and planning. We will conclude this chapter with general advice about ensuring competency.

BACKGROUND

The desire to ensure the competency of GIS professionals is a noble and laudable objective. It is also a necessary goal so as to guarantee the credibility of the GIS and the individuals who call themselves GIS professionals. This issue is therefore significant to the GIS community. Moreover, competency is required to maintain the credibility of the organization adopting the GIS.

It would be false to assume that because the GIS community currently does not have in place either a certification process or a system of accreditation that there is no current mechanism for ensuring the competency of professionals in the field. The current mechanism is basically an honor system that relies on both formal and informal networks of individuals and organizations to promote an atmosphere of integrity and ethical behavior. A number of well-respected individuals who have been instrumental in the creation and development of the GIS set a tone for the community as a whole. In addition, several professional organizations have developed a close working relationship to sponsor the annual Geographic Information Systems/Land Information Systems (GIS/LIS) Conference, providing a more formal set of norms and conventions of professional behavior.

So far, this semiformal arrangement appears to be working. The GIS community is still relatively small (although it is growing), and peer pressure seems to be adequate at this time to promote ethical behavior. As GISs become better accepted, the GIS community will continue to grow in size. Eventually the current arrangement may prove inadequate. It is therefore wise for the GIS community to begin thinking about a more formal mechanism for maintaining the quality and credibility of professionals in the field. In the meantime, there are several steps that an organization can take to ensure the quality of the personnel hired.

ROOTS AND RATIONALE OF CERTIFICATION AND ACCREDITATION

To understand fully the concepts of certification and accreditation, we must begin by exploring their roots, which are found in the ideas of expertise and professionalism. Expertise and professionalism are closely related, but they are by no means synonymous. Much of what we know about these concepts is derived from Weber's theory of bureaucracy (1946).

Expertise

Webster's dictionary (1984) defines "expertise" as "specialized knowledge or skill: mastery." We can and do think of expertise in many aspects of life, from dog training to neurosurgery to GISs. The specialized knowledge or skill required to demonstrate expertise is unique to each field.

In addition to being field specific, expertise is inherently time specific. For example, leeches were once a common and acceptable course of treatment within the medical community. Today, lasers and ultrasound are important medical tools over which modern physicians must develop mastery if they are to be considered experts. Similarly, in the field of GIS, paper and pens have been supplemented with (and in some cases replaced by) computers and their software. As innovations disperse throughout a field, the specialized knowledge of that field shifts to include the innovations, as well as to eliminate obsolete techniques and ideas.

Not surprisingly, expertise forms the backbone of professionalism, which is possible within specific fields largely because of the unique combination of information and knowledge within each field. Fields that cannot achieve consensus on exactly what mix of specialized knowledge is required to demonstrate expertise in the field have been known to question their own validity as a profession. Normally, this occurs within professions with diverse missions and members. Public administration is one such field.

A small body of specialized knowledge is not necessarily an impediment to the evolution of a profession, particularly if that body of knowledge is important to others. For example, beauticians and cosmetologists have been very successful in defining their profession. This group has managed to put in place certification processes in many states, thereby gaining public acknowledgment of the expertise of members of the profession.

In addition to its role as the backbone of professionalism, expertise is often cited as a key consideration in the development of certification and accreditation processes. The development of these mechanisms is frequently perceived and promoted as a means of guaranteeing that people who claim to be experts within a particular field actually possess the necessary expertise to make this claim. Thus, certification of individuals and accreditation of educational programs are designed, in part, to set a standard of competency for the field (whatever it is). Indeed, this rationale is frequently mentioned as a driving force for development of a certification process within the GIS community.

While the development of expertise may serve positive purposes (e.g., setting a standard of competency), Cayer and Weschler (1988) have noted that the expertise of professions and the concomitant control over information may lead to a concentration of power within professions (45). Similarly, Habermas (1970) suggested that experts may use their specialized knowledge to build a technocracy, thus gaining hegemony within their field. Weber (1968) likewise raised concerns about the elevation of technical experts to the status of a mandarin caste. There is a thin line between the concentration of expertise necessary to assure competency within a field and the use of expertise to create a technocracy. Given the technical nature of GISs, as members of the GIS community, we should be concerned about the creation of a GIS technocracy.

Professionalism

Webster's dictionary (1984) defines a "profession" as "a calling requiring specialized knowledge and often long and intensive academic preparation; a principal calling, vocation, or employment; [and] the whole body of persons engaged in a calling." This definition is consistent with notions of professions and professionalism embedded within Weber's theory of bureaucracy.

Weber and Professionalism

Writing at the turn of the century, Weber (1946) described the bureaucracy as an eminently enduring organizational model. A major reason behind the staying power of the bureaucracy is professionalism. Professionalism and professions are based on expertise, as noted above.

Weber further noted the importance of developing—and of guarding closely from outsiders—the body of knowledge or expertise that forms the foundation of a profession (1946, 233). Professions typically use education and sometimes certification examinations as a means of limiting entry into the professions. In addition, professional publications and networks facilitate the development and diffusion of a common language—sometimes better described as jargon. This shared language serves a valuable function: it helps to identify who is a member of the profession and who is not, and may be used deliberately to make entry into the profession more difficult.

While the development of expertise may serve positive purposes (e.g., setting a standard of competency), Weber (1946) raised concerns about the elevation of technical experts to the status of a mandarin caste. He noted that many professions gain a virtual monopoly in their area of expertise, which makes it very difficult for outsiders to evaluate the performance of members of the profession. The medical profession is a prime example (Berlant, 1975).

Pugh's Six Characteristics of a Profession

Building on Weber's assertion that expertise and the protection of that expertise are important prerequisites for the development of a profession, Pugh (1989) identified six characteristics of a profession. Most of these characteristics are self-evident; others need some explanation. The first of these characteristics is a cast of mind, or a self-awareness. The second trait of a profession is the possession of a unique body of knowledge necessary for the performance of professional duties, consistent with Weber's notion of expertise.

As the profession coalesces, it develops a third trait, what Pugh called "a social ideal to unify those working within an occupation." As an example, Pugh suggested that "for public administration, the consolidating vision was a knowledgeable, responsible, and proficient public service, the humane and efficient promotion of the common defense and general welfare, and the promotion of democratic institutions" (1989, 2). We may think of this characteristic as a professional culture.

Eventually, as the profession evolves, members of the professional community join together formally to create a professional organization—the fourth characteristic of a profession. Frequently, professional organizations establish a journal, newsletter, electronic mail network, or some other mechanism (or mechanisms) for communication among members.

These publications and networks become integral means for continuing the development and maintenance of the profession's expertise, as members share new ideas and refine (and sometimes eliminate) old ones. In addition, these publications and networks facilitate the development of professional jargon, which serves a valuable function by helping to identify who is a member and who is not. Jargon may be used deliberately to make entry into the profession more difficult.

According to Pugh, the fifth trait of a profession is "a hall of fame, a gallery of luminaries" (1989, 3). Individuals become part of this hall of fame by performing works in support of the profession, including theoretical and scholarly contributions, teaching and mentoring activities, and general advocacy on behalf of the profession.

Finally, a mature profession has a code of ethics. A code of ethics implies that the profession not only takes responsibility for a standard of competency among practitioners, but it endeavors to assure that its members will use their expertise ethically at all times. Professions may adopt any of several mechanisms to encourage ethical practice, including peer pressure and sanctions such as fines, suspensions, or even expulsion from the profession.

A Unified Model of a Profession

By including the essential elements of a profession as described by Weber and Pugh and combining similar or overlapping characteristics, it is possible to identify five key elements of a profession:

1. The existence and growth of a unique body of knowledge (expertise).
2. The rise of a professional organization.
3. The evolution of a shared language.
4. The development of a professional culture and lore (including a "hall of fame").
5. Shared concerns that form the basis for a code of ethics.

Using these criteria as the basis, there appears to be significant evidence supporting the notion that expert knowledge of and proficiency in the use of a GIS constitutes a profession (or at least is well on its way to becoming one).

Evidence That There Is a GIS Profession

There is ample evidence to suggest that the GIS community is well on its way to becoming a profession. As we will show, the community meets the first four criteria fully, and has the rudiments of the fifth characteristic as well.

Existence and Growth of a Unique Body of Knowledge (Expertise). Professional expertise can be found in two separate areas: research and teaching. Generally, expertise is maintained and shared through the written (or, more recently, the electronically transmitted) word. The GIS community has a growing body of expertise, both in research about GISs and in the teaching of knowledge about and use of the GISs.

Much of the early research on GIS exists in gray or fugitive literature, such as proceedings of professional meetings. There is a growing concern within the GIS community that this literature be cataloged and available for current and future researchers in the field. This suggests that the GIS community acknowledges the existence and importance of GIS expertise and is eager to develop and preserve it.

As GISs have evolved, the literature in the field has become easier to find because it is available in more mainstream sources. For example, there is a growing list of texts and collected readings on GISs, including books by Aronoff (1989), Burroughs (1986/1989), Huxhold (1991), and others. The recently published GIS reference *Geographical Information Systems: Principles and Applications* (Maguire, Goodchild, & Rhind, 1991) is a prime example and source of expertise in the field.

In addition, articles on GISs may increasingly be found in a variety of scholarly journals in fields such as geography, urban planning, landscape architecture, and surveying. Moreover, the creation of the *International Journal of GIS* is devoted to sharing and maintaining GIS expertise.

The development of the National Center for Geographic Information and Analysis Core Curriculum represents another incubator and storehouse of GIS expertise. The Core Curriculum has become a mainstay for GIS education in the United States.

Rise of a Professional Organization. Evidence that the GIS community is well organized can be found in the success of the GIS/LIS conference in the United States as well as the annual European Geographic Information Systems conference. In the United States, GIS/LIS is cosponsored by five separate organizations: the Association of American Geographers (AAG), the American Congress on Surveying and Mapping (ACSM), Automated Mapping/Facilities Management (AM/FM) International, the American Society for Photogrammetry and Remote Sensing (ASPRS), and the Urban and Regional Information Systems Association (URISA). GIS specialty groups exist within these organizations as well. The National Center for Geographic Information and Analysis, whose mission is to perform and promote research on basic issues regarding GISs, is another organization in the field.

When we use the phrase "the GIS community," it refers to the organizations listed above. Although the GIS community has not yet formally established an entirely new organization devoted solely to promoting GISs, we have a very well established coalition comprising members of several organizations. The professional organization of the GIS community can be shown graphically (see figure 9.1).

Although the current arrangement is an amalgam of several organizations rather than a single formal professional organization, the ability of the consortium to organize the GIS/LIS conference year after year suggests that this amalgam is effective. There has been talk about creating a formal GIS organization; but thus far, the talk has not turned to action. Even in the absence of such a formalized arrangement, it is safe to say that the GIS community is well organized. The effectiveness of the current arrangement is further evidenced by the existence of the *International Journal of GIS, GIS World,* the GIS list on Bitnet/Internet (an electronic mail network that connects people who are interested in GIS), and other publications and networks that enable GIS professionals to communicate readily among themselves.

Evolution of a Shared Language. The development of expertise in GIS, along with the coalescence of the GIS community as an effective, functioning group has promoted the evolution of a shared language. The GIS community speaks a jargon unto itself.

For example, when GIS professionals say "GIS," they mean "geographic information system" (not, e.g., "guidance information

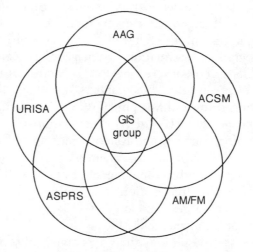

FIGURE 9.1. The GIS group amalgam.

system," which is a real computerized system used by high school guidance counselors). When they mention "DIME," they understand that this means "dual independent map encoding," a system used by the U.S. Bureau of the Census that is a predecessor of the current "TIGER" (topologically integrated geographic encoding referencing) files. However, members of the GIS community would never confuse the "TIGER" files with a large member of the feline family.

Similarly, GIS professionals readily throw around phrases such as "object-oriented," use acronyms like "DLG" (digital line graph), and discuss a variety of proprietary geographic information systems, including "Arc/Info," "Idrisi," "Intergraph," "GRASS," "SPANS," and others.

A discussion among members of the GIS community would make little sense to an outsider, both because of the technical nature of GIS as well as because of the shared language that the community has evolved.

Development of a Professional Culture and Lore, Including a Hall of Fame. The notion that a profession develops its own culture and lore is central to the creation of a distinct professional image. In this context, the professional culture is expressed through networks of GIS managers and scholars, the mentoring process that often exists within and across organizations (including universities), and the celebration of important milestones in the profession. We can identify several important milestones in the development of GISs. For example, the development of the Canadian GIS, considered by many in the community to have been the first true GIS, is a major milestone. Similarly, the adoption by the United States Census Bureau of the GBF/DIME files represents an important watershed. Not surprisingly, the GIS community in the U.S. has also eagerly anticipated the advent of the TIGER files.

The idea of a "lore" refers to the collection of myths, stories, and a Hall of Fame comprising early pioneers in the field. GIS users frequently speak among themselves of the value and benefits of GISs, firm in the belief that GISs can help improve decision making in both the public and private sectors. An important mission within the community is promoting the adoption of GISs for a wide variety of applications. In general, GIS professionals have been quite successful in pursuing this mission, as shown by the proliferation of GISs in a growing variety of applications. Further, much like the manner in which IBM employees revere the memory of Thomas Watson, Sr., or

Disney employees look back to the impact of Walt Disney on their company, the GIS profession is in the process of developing its own hall of fame made up of some of the early researchers and technical giants in the field. People such as Roger Tomlinson, Ron Abler, Jack Dangermond, Robert Aangeenbrug, Duane Marble, Mike Goodchild, and others figure prominently in the development and evolution of the GIS field into an identifiable profession.

Shared Concerns and the Formation of a Code of Ethics. As the GIS field grows and proliferates, many of the teething problems emerge, as they are bound to do with the introduction of any new technology. Further, there is the need to address some of the unforeseen side effects of the technology, as in the legal problems resulting from cases of rights of privacy versus expanded access to information. Pugh has made the point that it is usually at this stage that some dialogue on shared concerns begins to make itself heard as GIS professionals attempt to establish a set of rules of behavior, including some form of a code of ethics. Although the GIS community does not currently have a formal professional code of ethics, members frequently discuss issues that may be viewed as precursors to the development of such a code. For example, recent discussions about the possible need for certification of GIS practitioners have focused on the need to ensure competency within the field. In addition, issues related to data quality arise every time members of the GIS community assemble. Moreover, the November 1993 workshop on Geographic Information and Society focused on societal problems that may accompany the widespread diffusion of GIS. These three items alone suggest that GIS professionals are concerned about the quality of their professional contribution to the larger society.

However, the concern of GIS professionals does not stop with mere competency. Among the issues that these professionals have raised, questions about policies for pricing data and information made available by GISs is a critical ethical concern. Pricing policies are of particular concern when the databases under consideration are gathered by using public funds and the purchaser plans to use the data for private economic gain.

In addition, Smith (1992) cut to the heart of ethics and morality in the use of GISs by questioning GIS military applications, which, he noted, have driven the very development of the technology. Smith pointed out that "GIS and related technologies" contributed to "the killing fields of the Iraqi desert" (257). Smith raised the concern that the ability of the GIS to provide computerized representations of real-world military death and destruction may make it even easier for us to detach ourselves from the often horrific outcomes of international military conflict. This is a valid concern, and one which bears etching in our consciences. However, Smith himself pointed out that geography and geographers have long and often been at the heart of international conflicts and conquests: his very loose interpretation from a French quotation becomes, "Geography, thy name is war" (259). In light of the historical link between geographers and war, targeting GIS as a negative force in international conflicts must not occur at the expense of letting geographers off the hook for their own individual ethical and moral lapses.

Smith's general point, that is to say, that the implementation of geographic information systems is not a value-neutral endeavor, is echoed by a growing chorus within the geographic community. Researchers such as Dobson (1993), Goodchild (1993), Pickles (1993), and Sheppard (1993) emphasize that a Pandora's box of societal repercussions, notwithstanding those that are beneficial, are bound to accompany the widespread adoption of GIS, particularly in the public sector. These issues are covered in greater detail in Chapter Eleven.

This brings us back to the point that the development of a code of ethics is necessary for the GIS community. As individuals, GIS professionals must first accept the enormous capabilities that GIS brings and must take individual responsibility for their own actions. However, it is also necessary for them to develop the parameters of ethical behavior for the profession in order to prevent, as far as possible, the unethical or immoral behavior of others. These shared concerns—and others—form the foundation for a code of ethics within the GIS community.

GIS: A New Profession

It appears, based on the description above, that expert knowledge and proficient use of GISs can justifiably be called a new profession. There can be no doubt that the community of GIS users has expertise, a shared language, as well as a professional culture and lore (including candidates for a GIS Hall of Fame).

The cohesiveness and effectiveness of the consortium of GIS/LIS-sponsoring organizations forms the basis for a GIS professional organization. We may quibble about this point and argue that since the field lacks a true, unified professional organization owing to the informality of the GIS group, it does not qualify as a profession; however, the reality is that GIS scholars and practitioners function as a professional organization in spite of the informality of the current arrangement. The fact that those knowledgeable about and proficient in the use of GISs have organized several major conference series and created a professional journal and electronic mail network speaks for itself.

Similarly, while it is true that GIS professionals officially lack a code of ethics, ongoing discussions have laid the foundation for the eventual development of such a code. The enduring nature of several of the debates suggests that there is great consensus among GIS professionals regarding areas that might be covered by a code of ethics. Moreover, the persistence of these issues suggests a need for some resolution, which could come in the form of a code of ethics.

The evolution of GIS usage into a profession has helped to stimulate interest in assuring the competency of GIS professionals. Initially, certification was identified as the means to achieve this worthy end.

CERTIFICATION IN OTHER FIELDS

The idea of certification is not new. Indeed, many professions—including medicine, law, surveying, planning, and cosmetology—have long-established certification procedures. These certification processes are characterized by varying degrees of effectiveness. In this section we will look at some of the successes and failures of two of these processes as a means of providing insight into certification in GIS.

Although preceding descriptions have implied that the concepts of professionalism and certification are internal and self-regulatory as well as self-promotional processes, it is important to make this aspect explicit. It is within this context that we understand the effectiveness of the certification process. Effectiveness may be defined in two ways: first, it may refer to the appropriate and ethical behavior of members of a profession in meeting their responsibilities to the larger society; second, effectiveness may refer to the success of members of the profession in the marketplace, and the ability of the profession to ensure that the practitioners it has certified are hired in preference to practitioners who do not have the profession's official seal of approval.

An examination of certification in the medical and planning professions helps to illustrate these points.

Development of a Professional Monopoly: The American Medical Association

One of the most successful of all professional organizations is the American Medical Association (AMA). Formed in 1847, its raison d'être was to upgrade the educational standards in medicine (Berlant, 1975, 226). At the urging of Nathan Smith Davis, whose ideas formed the basis for early AMA policy, the AMA adopted a system that separated teaching and licensing. Medical schools continued to function as they always had, while licensing was instituted at the state level, with medical societies having responsibility for appointing individuals to the state licensing boards. According to Berlant, the AMA established a policy that required *both* a diploma and a license in order to enter the practice of medicine; previously *either* a diploma from a medical school or a license had been sufficient to allow doctors to set up a practice.

"Protection of the public against quacks," Berlant noted, was the primary justification for the development of what eventually became a monopoly in medical services by the AMA (227). Following in the German tradition of state licensing superimposed on university examinations, Davis effectively promoted legislation that ultimately pressured medical schools into a specific line of development—what Berlant described as orthodox medicine (227). The imposition of this legislation effectively made orthodox medicine (i.e., that promoted by the AMA) the only medicine that would receive the legislature's official seal of approval. Orthodox medicine effectively became the basis for accreditation of medical programs as well as the basis for certification examinations. According to Berlant, what had previously been professional dominance by the AMA became hegemony; with the support of state legislatures, this hegemony became a virtual monopoly.

In the years after its establishment, the AMA became synonymous with quality health care. Practitioners and products alike proudly displayed the AMA's seal of approval. In recent years, however, concern about the monopolistic domination by the AMA has arisen. Complaints about the arrogance of doctors have become commonplace. Until very recently, one dared not question the medical judgment of the physician: he or she alone possessed the medical expertise to save a life. The AMA had come to resemble the technocracy described by Habermas (1970). Berlant suggested that the legislative seal of approval played an important role in the evolution of the AMA into an exceedingly powerful monopoly. We might also conclude that the complexity of medicine itself, along with the universal importance of medical care, were crucial factors in the evolution of the AMA.

Voluntary Certification: The American Institute of Certified Planners

If the AMA stands out as a professional organization whose certification process assured a virtual monopoly for its members, then the American Institute of Certified Planners (AICP) is an example of what happens when certification is a totally voluntary exercise.

In his discussion of the AICP certification examination, Rasmussen (1986) argued that "certification by the professional society means planners themselves control their own occupational standards. In contrast, state licensing or registration gives the state legislatures statutory control over occupational standards by regulating the right to practice planning or to use the title of planner" (7). Rasmussen seems to imply that the current arrangement promotes maximum autonomy for the AICP. However, comparing the AICP to the AMA suggests that there is a price for this autonomy.

As described above, the AMA exercises considerable authority over the appointment of members to state medical licensure boards. This authority, Berlant argued, has resulted in the development of a state-approved virtual monopoly by the AMA over the entire medical practice. One aspect of this monopoly is that only physicians certified by the AMA may practice medicine. By contrast, the planning profession is open to any individual who cares to call himself or herself a planner.

While certain jobs in planning (most commonly, academic positions) require that candidates have AICP certification, most do not. Many do not even require completion of a degree program (either accredited or unaccredited) in planning. For example, in the June 15, 1991 issue of the American Planning Association's *JobMart*, of 37 positions advertised, none required AICP certification, while only one gave preference to AICP planners. Furthermore, barely half (18) required a degree in planning, and 10 listed planning as one of several acceptable degrees. The nine remaining positions did not specify educational requirements or did not specify planning among their lists of acceptable programs of study.

From the perspective of the planning professional, the value of AICP certification seems limited, at best—at least as far as employment prospects are concerned. Furthermore, one could logically challenge Rasmussen's implication that the AICP exercises autonomous control over the quality of planning practitioners. In fact, if one need not achieve AICP certification in order to work as a planner—and if, indeed, *most* planning positions do not require AICP certification— then it is clear that the AICP exercises almost no quality control in the planning marketplace. Rasmussen himself admitted that only about 20% of the members of the American Planning Association are also members of the AICP (and, by implication, have passed its certification examination). Presumably most members of the American Planning Association are professional planners.

The AICP certification procedure is intended to ensure quality among planning practitioners. It is probable that planners with AICP certification possess high-quality skills and experience. However, in practice, the AICP exercises little control over the practice of planning in the United States. Without some mechanism (either voluntary or coercive) to limit entry into planning practice exclusively to AICP members, the profession will be unable to ensure quality across the board.

The differences between the AMA, which has a virtual monopoly on medical practice in the United States, and the AICP are readily apparent. In the first instance, their approaches are very different: the AMA assiduously courted legislators in its early stages to ensure the dominance of its practice of medicine over all other types. The AICP, by contrast, has resisted what it perceives as the threat of legislative control of its profession. The outcomes are dramatically different: the AMA has achieved a stranglehold on medical care in the United States; the AICP is a lofty goal to which many planners may aspire, but which few actually attain.

ACCREDITATION: A BRIEF DESCRIPTION

Accreditation differs from certification in that its focus is at the institutional level rather than the individual level. Accreditation has a long history in academia, with colleges and universities earning accreditation as a whole and academic programs earning accreditation within their specific professional or academic sphere. In the case of GISs, the latter approach is relevant.

Generally, the first step in developing an accreditation process is the establishment of a specific set of evaluation criteria. These criteria generally include standards for personnel (both quantity and quality), facilities and equipment, and coursework. They are normally set by an organization that has credibility within the profession. For example, the AICP sets the standards for accredited planning programs at the university level. The National Association of Schools of Public Affairs and Administration sets the standards for accredited programs in public affairs and administration.

Once evaluation criteria and standards are in place, programs apply for accreditation, initially providing written documentation as required by the accrediting organization. If the initial application and materials are approved, the accrediting organization sends a site inspection team to examine the premises of the program requesting accreditation. Providing that the inspection team is satisfied that the program (including facilities, equipment and so on) meets or surpasses the evaluation standards, the program receives its accreditation for a predetermined period of time. When that period of time expires, the program must begin the process anew. However, if in the meantime, conditions within the program change in areas covered by the evaluation standards, the program risks losing its accreditation. This has been occurring in planning programs, which are required to have at least five full-time equivalent (FTE) faculty members to achieve and maintain AICP accreditation. In the current climate of university downsizing, faculty loss is becoming more common, jeopardizing some planning programs on the basis of too few faculty members.

Like certification, implementation of accreditation procedures runs the risk of encouraging a technocracy, and, therefore, must be carefully considered. If GIS professionals decide that a formal procedure for ensuring competency is desirable in their GIS profession, accreditation may have some advantages over certification. First, because there are fewer colleges, universities, and private providers offering GIS courses and programs of study than there are individuals who would aspire to become GIS professionals, there would be fewer schools seeking accreditation than individuals seeking certification. In a similar vein, the familiarity of GIS professionals with one another and with one another's programs may facilitate accreditation reviews.

For both accreditation and certification, dealing with the varied applications of GISs promises to be a challenge. Several challenges are discussed in greater detail below.

IMPLEMENTING CERTIFICATION OR ACCREDITATION

Implementing a certification or accreditation procedure for GISs would not be a simple task. It is not clear, based in part on the AICP's experience with certification, if it is necessarily even a worthwhile task. And in spite of efforts among surveyors to make GIS certification mandatory, the talk about certification has subsided. There are several reasons for this.

First and foremost, GIS professionals have simply not agreed that certification is necessary or desirable. Given the breadth of the community of GIS professionals, the inability to reach agreement should not be surprising. In order for certification to become a reality, several specific questions would need resolution:

1. Who will develop the examination or review criteria?
2. How will the examination/review approach the diverse applications of GISs?
3. What are the core skills and knowledge necessary to become a competent GIS professional?
4. How will certification examination or accreditation assure high standards among GIS professionals? Will the state become involved, as in the case of the AMA? Or will certification become voluntary?

It is clear the stated goal of assuring that GIS professionals will be fully qualified for their assignments is worthwhile. How the community of GIS professionals chooses to achieve this goal is another question entirely. In the foreseeable future, neither general certification nor accreditation is likely to become a reality.

ALTERNATIVES TO A CERTIFICATION EXAM

If the goal is to assure that GIS professionals will meet a minimum standard of qualification, then there are several ways to achieve this goal that may be used alone or in combination with one or more other alternatives. Among these alternatives is for the community to continue as it now exists, within an informal yet relatively cohesive network of GIS professionals. This framework, similar to formal professional organizations, relies on peer pressure as a means of guaranteeing appropriate and ethical professional behavior. GIS professionals keep each other honest.

It is also possible to formalize the current arrangement. Establishing a formal professional organization will make it increasingly possible to impose sanctions on members whose professional behavior is inconsistent with community norms and conventions. The disadvantage of this alternative is that most GIS professionals are already dues-paying members in at least one or two other professional organizations. The cost factor alone may cause many to avoid membership in a new GIS organization.

Another alternative to the institution of a certification exam would be to develop an accreditation system for GIS programs. Possibly using the National Center for Geographic Information and Analysis Core Curriculum as a starting point, GIS professionals could develop a set of performance criteria that GIS programs would be required to meet in order to earn accreditation. There are a couple of advantages in instituting accreditation at the school level (as opposed to the individual level). First, there are fewer schools than individuals, making the latter an easier target. In addition, many faculty members involved in GIS programs are already active GIS professionals. Reaching them will likely be easier than reaching individuals who come from a wider variety of backgrounds.

One final alternative to a certification exam is the development of a code of professional ethics for GIS practitioners. This alternative has the advantage of being completely voluntary. Moreover, it is the kind of action that recalls Mom and apple pie, and, therefore, should be relatively popular within the community.

WHAT TO DO IN THE MEANTIME

We conclude this chapter by reminding the reader of the original point: there is a growing need to guarantee the competency of GIS professionals. This point will certainly not be lost on anyone who has had to hire someone to help implement their GIS. There are several things that potential employers can do to ensure the hiring of qualified GIS professionals.

First, the potential employer should read as much as possible about

GISs and their specific applications to the employer's business. This will help the employer understand the business's needs more clearly and to identify the skills and capabilities required. Moreover, it will help him or her identify organizations that have accomplished similar implementations with which he or she might network. Reading will also help the potential employer learn more information about education in GISs that is available at various institutions of higher learning.

Second, the potential employer should network. Before hiring anyone, the owner(s) or management should attend local, regional, and, if possible, national GIS conferences and workshops. This will give them an opportunity to meet potential employees, either directly or through contacts they might make. In many cases, formalized procedures enable prospective employers and prospective employees to meet for face-to-face interviews during the course of the conference.

As part of networking activities, the potential employer should talk with other participants, the exhibitors, and presenters, about his or her needs, and then solicit their advice. These conversations may identify specific people who would be right for the job opening or, at the very least, may help the employer identify the characteristics you should be looking for. This will help you to develop a position description. These networking activities should also help the employer identify specific avenues by which the job opening can be advertised.

Third, the employer must remember that vendors are a valuable resource. If an employer has already chosen a GIS, the vendor can be of particular help at this phase. Vendors have a powerful economic interest in the successful implementation of their GIS. The vendor most likely has formalized training programs (in house or sometimes on site) that potential employees should have completed. In these situations, the person who has successfully completed such a training program will often have earned a certificate of completion from the GIS vendor.

In addition, vendors also tend to have extensive contacts in business, in the public sector, and in academia. They may be aware of programs that produce GIS professionals that fit an employer's specific needs. They may have heard of someone who is looking for a new professional challenge. Vendors are an important link in any network.

CONCLUSIONS

As noted, it would be erroneous to assume that the community of GIS professionals currently has no system in place to guarantee the competency of its members. While no formalized system such as certification or accreditation yet exists (and may never exist), the community is still small enough that members tend to know or know of a large proportion of the active participants in the community. This has contributed to the development of a relatively cohesive group that operates largely on the basis of peer review. So far, this format has been adequate. To the best of our knowledge, there have been no GIS scandals that have damaged the credibility of either the technology or the profession.

On the other hand, the rapid pace at which organizations are adopting GIS means that we must not become complacent. The demand for more and more professionals means that finding qualified professionals may become more difficult in the short run. In the long run, we can expect market forces to begin to blunt any shortages. Increasingly, institutions of higher learning are implementing GIS courses and programs that will produce more professionals. At the same time, the technology continues to become easier to use. This itself will help ameliorate any employee shortages.

Organizations seeking to employ GIS professionals should take the usual precautions that they would take in hiring any employee. Verify resumés, check references, and hold careful interviews (multiple interviews, if need be). These practices, along with those described earlier, will make the difference. In addition, however, organizations will also benefit from gaining a thorough knowledge of GISs and the needs that these systems serve.

Policy Conflicts and the Role of GISs

One of the claims made by advocates of GISs is that these systems can help to minimize conflicts over land use by providing more and better (more accurate) information about the subject of the conflict. This claim overlooks an important source of conflict: the underlying value differences represented by conflicting parties. In this chapter we will explore the role of GISs in the public policy sphere and will suggest that GISs will tend initially to increase rather than decrease conflict, since geographic information and analyses made possible by GISs can be used selectively by conflicting parties to support their positions. However, this conflict can be viewed as a positive feature in a democracy, since it represents open dialogue concerning differences of opinion that must be fully explored as a precondition for acceptable public policy resolutions.

In addition to their tremendous potential within private organizations, GISs are becoming increasingly important tools in the public sector. Although the primary orientation of GISs to date has been toward information and infrastructure management rather than spatial and policy analysis (Goodchild & Getis, 1991), they are growing in importance as tools for analysis and public-policy development, including such tasks as land-use analysis and political redistricting. It has been suggested that GISs may play a role in minimizing conflicts among competing interests regarding land-use issues by making available more and better information. Better information is usually construed to mean more current and accurate information. It is hypothesized that this information will improve analysis and facilitate agreement among the competing parties. This suggestion appears to be somewhat naive, since it overlooks the underlying value conflicts that precipitated the initial conflict in the first place.

In this chapter we will explore the likelihood that GISs can minimize conflicts and will suggest (1) that the ready access to information made possible by the proliferation of GISs will lead to *increased*, not decreased, conflict in the short run, as a greater number of diverse interests harness these powerful tools in support of their objectives, but (2) that eventually the conflict will level off, although at a higher level than previously existed.

The logic behind this suggestion lies in research that identifies two sources of conflict: disagreement on facts (cognitive conflict) and disagreement regarding values (interest conflict). While GISs can influence facts in a particular conflict, there is no reason to expect that they will do anything to mesh competing values. Value conflict, therefore, will remain, regardless of the amount of information gathered to resolve it. At the same time, more information increases the number of "facts" that can become the basis for further conflict.

This chapter will begin with a discussion of the theoretical underpinnings of cognitive versus interest conflict. A discussion will follow about how cognitive and interest conflict might respond to the addition of GISs as analytical tools, taking as an example the conflict created by the proposed siting of a nuclear generating facility in northeastern Indiana in the 1970s. The chapter will conclude with a discussion of the value of conflict in a democracy and of the role of GISs in that conflict.

COGNITIVE VERSUS INTEREST CONFLICT

Among the potential benefits of implementing GISs, the question has been raised: Can GISs minimize conflicts regarding land use? The belief that GISs will help minimize conflict begins with the premise that GISs make more readily available and accessible greater quantities of data and information within the framework of a computerized package for analysis of the data. The belief, therefore, rests on two basic assumptions: (1) that more information is necessarily better, and (2) that all participants in a conflict will agree on the validity of both the data and the decision models used within the framework of GISs. In short, this view assumes that there is an objective reality on which all parties can agree. Our contention—one that is consistent with recent literature on conflicts in policy making as well as with earlier theoretical work of Weber (1946, 1968) and Habermas (1981)—is that these assumptions are inaccurate.

Weber challenged the assumption that there is an objective reality on which people can agree, basing his argument on the existence of differing (and often competing) values that people hold near and dear. Debates that have values at their core represent a source of irreconcilable conflict, according to Weber. No amount of rational appeal is guaranteed to sway an individual from his or her values. In fact, argumentation based on values—that is, substantive rationality— is often futile. In Weber's view, "Scientific pleading is meaningless in principle because the various value spheres of the world stand in irreconcilable conflict with each other" (1946, 147).

Weber (1968) described argumentation based on values as being substantially "ends rational." By ends-rational action, Weber referred to the goal choices of individuals, closely intertwined with individually held values. Such actions hold intrinsic value for the actor, regardless of the outcome, and are ends in and of themselves. For example, pro-life activists demonstrating outside abortion clinics and intimidating women seeking abortions undertake such activities as a personal mission. While they may stop a few abortions, their actions do little to sway public policy makers to outlaw abortion or otherwise put an end to abortion; their aggressive tactics may even hurt their cause. Still, in their minds, they are doing the right thing, regardless of the long-term outcome. The protests themselves, therefore, have intrinsic value.

In contrast, Weber (1968) discussed the notion of formal rationality, which is "means rational." By means-rational action, Weber referred to the strategy that an individual follows to achieve his or her goal. The means-rational person chooses a strategy that has the greatest chance of achieving his or her ultimate goal and is willing to forgo short-term gains for long-term results. Since values are not constantly at the forefront for the means-rational person, discussion and persuasion are likely to be more successful in achieving resolution.

Again, to offer an example, consider the case of a manager seeking a promotion. In an effort to maintain a positive image with his or her coworkers and superiors, this individual may make a point of avoiding engaging in any conflict that could potentially threaten the chances of a promotion. In effect, the manager is willing to avoid or even lose short-term conflicts that may hinder the long-term likelihood of success. The manager has made a conscious decision to separate the lure of relatively valueless short-term gains from the valued long-term goal.

Brubaker (1984), critiquing Weber, commented that "formal rationality is a matter of fact, substantive rationality a matter of value" (36). Policy decisions are almost always value laden; hence, the use of GISs in land-use analysis and policy development would do nothing to minimize the divergence of values held by the participants in the policy process but, instead, would only insert new tools into that process. At most, GISs would make more information available more rapidly. The addition of more facts at a more rapid pace may, in fact, only lead to increased conflict.

In contrast to Weber, Habermas has emphasized the existence of an objective reality and has been less concerned about values. According to Habermas (1968), in conflicts regarding this objective reality, the more rational argument—that which raises greater validity claims—should hold sway (99). Habermas emphasized the role of issue definition in disagreements among interested parties:

> A definition of the situation by another party that prima facie diverges from one's own presents a problem of a peculiar sort; for in cooperative processes of interpretation no participant has a monopoly on correct interpretation. For both parties the interpretive task consists in incorporating the other's interpretation of the situation into one's own in such a way that in the revised version "his" external world and "my" external world can—against the background of "our" lifeworld—be relativized in relation to "the" world, and the divergent situation definitions can be brought to coincide sufficiently. (100)

Rational communicative action would be the means to accomplish this task and ultimately result in agreement. This agreement will then form the foundation for our later argument regarding the resolution of the issue itself. According to Habermas, the participants will ultimately settle their disagreement by presenting their arguments, with the more rational argument winning.

Habermas stressed the importance of developing a framework of social norms among the parties engaged in discussion. His concept of communicative action presumes an objective reality about which meaningful argument can occur, permitting the participants to reach a logic-induced understanding. Weber's concept of substantive rationality is directly at odds with Habermas's concept of rationality as defined by the latter's theory of communicative action. First,

Habermas's theory assumes that reaching understanding is an important goal of human association. Weber (1968), on the other hand, recognized that "rational exchange is only possible when both parties expect to profit from it or when one is under compulsion because of his own need or the other's economic power" (73). "Not ideas, but material and ideal interests, directly govern men's conduct" (1946, 280). In human terms, Weber's idea holds that most people engage in conflict for personal as well as professional reasons and that self-interest is often the driving force behind these conflicts and their possible resolution. Habermas seems to have implicitly assumed that when all the facts are in, the logical decision will hold sway since those in the wrong will see the error in their viewpoints and will capitulate to some objective "truth." Weber took a more realistic (socially-driven) approach, which suggests that people often invest far too much emotional capital into their positions to surrender gracefully in the face of opposing argument, confrontation, or even direct logical refutation of their views.

Against this theoretical backdrop, more recent studies of land-use decision making pick up the theme of ideas (facts) versus interests (values), and suggest that science and technology alone will not minimize conflicts. In their analysis of land-use studies using multiobjective programming models, Wang and Stough (1986) discovered that most such models fail to consider the cognitive processes of decision makers and noted the importance of cognitive conflict within the policy-making process. According to these authors, "Cognitive conflict exists when individuals base their decisions on different facts or on the same facts construed differently" (107). By contrast, interest conflict occurs when individuals have different values or desire different outcomes. Belief in the ability of GISs to minimize conflict rests on an implicit assumption that the participants in the decision-making process will agree on the relative importance of the facts (data) and that all participants will construe those facts identically.

Other authors (Brill, Flach, Hopkins, & Ranjithan, 1990; Stough & Whittington, 1985) have responded to the problem of interest conflicts with the development of multiobjective models that suggest an array of resolutions (as opposed to a single "best" resolution) to land-use conflicts. This strategy may prove more hopeful as a conflict-resolution technique and may offer a constructive role for GISs, which may be successfully used to develop alternative scenarios that may facilitate compromise among competing interests.

A MODEL OF CONFLICT

The notion of two very different types of conflict, then, is well accepted. If we accept this idea, we can easily envision a situation wherein any given debate over land use will have two components: interest (value) conflict and cognitive (fact) conflict. Since many such debates are long lasting, we can further imagine an ongoing conflict over time, with interest (value) conflict remaining more stable than cognitive (fact) conflict. Figure 10.1 illustrates this model.

AN EXAMPLE OF CONFLICT

A real-life example that illustrates cognitive and interest conflict may be drawn from the debate over the proposed siting of a nuclear generating facility adjacent to the Indiana Dunes National Lakeshore, in northwestern Indiana (Obermeyer, 1990a). First proposed in 1972 by the Northern Indiana Public Service Company (NIPSCO), the Bailly plant, as it was called, was designed to provide energy to homes, business, and industry (including the steel industry) in the region. Its timing seemed ideal, at least initially—coinciding with the energy fears generated by the "oil shocks" of the early 1970s.

However, neighbors of the proposed site had another opinion. A group of homeowners in the nearby residential areas of Ogden Dunes, Dune Acres, and Beverly Shores, described as being "among America's

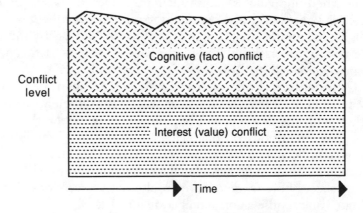

FIGURE 10.1. Routine conflict level before introduction of GIS.

finest exclusive communities" (Mayer, 1964), learned of the proposed facility and began what was to become a successful 10-year legal war to derail the project. These residents were unusual in at least two respects: as a group, they were both well-educated and wealthy. These characteristics were a distinct advantage to them as they fought their war, providing them with both the knowledge and the economic resources to do extensive legal battle with a major public utility. These people united and called themselves the "Save the Dunes Council" (Obermeyer, 1990a, 80).

The conflict at the Dunes involved several competing interests in addition to NIPSCO and the Save the Dunes Council. Environmentalists expressed concern about the Lake Michigan shore, the most extreme of them favoring a broad expansion of the park land. Had such an expansion been adopted, many of the members of the Save the Dunes Council would have lost their homes. The steel industry, a powerful interest group in 1972 (although weakening) was also a neighbor. While steel industry managers favored the facility because it promised more energy, workers opposed the plant because they feared the efficacy of evacuation plans. The region itself was undergoing substantial commercial development as well, centering on the intersection between Interstate 65 and State Route 30. Most of these groups eventually played a role in the conflict.

Historically, competing interests have consistently been at the heart of controversy surrounding the Dunes. Certainly, economic considerations have formed a major part of the controversy engulfing the Dunes; but Engel (1983) claimed that more personalized attachments to the Dunes—an almost spiritual link between the Dunes and those who love them—formed an important element of the controversies surrounding the local land-use issues. According to Engel, owing to this "spiritual" attachment, values were extremely relevant to any discussion of conflicting land use at the Dunes and in surrounding areas.

At the heart of the conflict surrounding the siting of the Bailly nuclear generating facility was a difference in values between the public utility and the Save the Dunes Council. NIPSCO's interest lay in its ability to provide an adequate supply of electricity for its residential and commercial customers. The Save the Dunes Council's interest lay in its desire to protect and preserve the residential communities (where the members of the council lived) from the intrusion of such a large facility and its consequences, both intended and unintended. Later, environmental groups and two local labor unions joined the council to promote their own interests, which were different from, though consistent with, those of the council. Although the council expressed its concern about the Dunes as an important environmental area, it is hard to imagine that the members were not more concerned about the effects—both real and potential—of the proposed facility on their homes and property values.

Ironically, both the council (and its allies) and NIPSCO used facts to support their positions in challenges, counterchallenges, and rebuttals brought before the Nuclear Regulatory Commission (NRC) and the courts. For example, the Save the Dunes Council and its allies, known as the joint intervenors, challenged NIPSCO's choice of a site for the Bailly facility and charged that the NRC had violated its own regulations regarding the acceptable distance between a nuclear generator and population centers. The intervenors argued that the NRC should have calculated the distance from the Bailly site to the nearest population center from the border of the city rather than from the city center.

Failing in their challenge in administrative hearings and appeals within the NRC, the joint intervenors took their case to the U.S. Court of Appeals, which ruled in their favor. An appeal from NIPSCO sent the case to the U.S. Supreme Court, which reversed the decision of the Court of Appeals, holding that the NRC was the official expert on nuclear energy and, therefore, that it alone was responsible for developing an appropriate method for calculating the distance between the facility and the nearest population center.

Later on, two local chapters of the United Steelworkers Union joined the Save the Dunes Council and the joint intervenors in the council's concern about the method for calculating the distance between the proposed nuclear facility and population centers. The concern of the Steelworkers Union was that NIPSCO's method for calculating the distance between the facility and the steel factories exaggerated the distance. The union was concerned about safety and the ability to evacuate the nearby factories in case of a nuclear accident. While the union agreed with the Save the Dunes Council that NIPSCO's distance calculations were flawed, their specific interest was neither in protecting the exclusive communities of the region, nor in saving the Dunes for conservation's sake: rather, they were concerned about their own personal safety during evacuation. Presumably, had NIPSCO succeeded in delivering an acceptable evacuation plan, the union would have been satisfied. These differences suggest that the union and the Save the Dunes Council agreed at the level of cognitive conflict but not at the level of value conflict. Still, to the extent that their agreement could enlarge the opposition against a common opponent, their unity was assured.

The joint intervenors (especially the environmental groups) and NIPSCO also disagreed about other "facts" in the case. The joint intervenors charged that NIPSCO's excavation was draining Cowles' Bog; NIPSCO disagreed. The intervenors alleged that NIPSCO had failed to consider other sites and other sources of energy as alternatives to the Bailly nuclear facility as required by the National Environmental Policy Act of 1969; again, NIPSCO disagreed. These and many other disagreements constituted a degree of conflict over the substantive (and administrative) facts in the case.

While we can separate the cognitive conflict from the interest conflict in the Bailly case on an intellectual basis, it is also clear that the two types of conflict are closely interrelated. Moreover, it appears in this case that the cognitive conflict was driven by the existence of interest conflict. Bluntly put, the joint intervenors used cognitive conflict as a way to promote their own interest—that is, the demise of Bailly. Again and again, over the course of nearly 10 years, the joint intervenors raised challenges based on facts in the case. Sometimes they won; more often they lost the battles. However, their persistence paid off in the end.

In 1982, 10 years after NIPSCO had first proposed the Bailly nuclear facility, it withdrew its proposal. By that time, NIPSCO had succeeded only in excavating a very large hole, at the cost of some 200 million dollars. Ultimately, the direct and indirect costs of addressing the challenges of the joint intervenors became overwhelming, and NIPSCO was forced to back down. Eventually, the public utility did build another electrical generating facility, a coal-fired plant in Wheatfield, Indiana. Construction of that facility went on without opposition.

A HYPOTHESIS ABOUT GISs AND CONFLICT

As this case study suggests, in practice as well as in theory, cognitive conflict and interest conflict are separate. However, this case also suggests that issues raised at the level of cognitive conflict may promote specific interests as well. We must keep in mind that the primary forum in which such conflicts are played out, the U.S. legal system, is an adversarial system. It is inherently conflictual.

Among other things, this case demonstrates the value of using specific cognitive issues to challenge and rebut opponents in the courts. Although the joint intervenors obviously preferred winning to losing—and a big win, either in the administrative legal environment of the NRC or in the U.S. judicial system itself would have meant the immediate and final demise of the Bailly facility—just playing the game advanced their cause. Ultimately, playing the game enabled them to win what turned out to be a game of attrition.

It is not difficult to imagine the value of GISs as a means of producing more information that may be used to support specific positions on a given land-use issue, especially in a public setting. The Freedom of Information Act assures open access to nearly all government-owned information. In addition, the government makes additional efforts to improve public access to certain types of information (e.g., the TIGER files). The result is that much of the georeferenced data that can be used in implementing GISs is readily available in a compatible format.

In addition, the price of GIS hardware and software is decreasing in real terms. As computers themselves become accepted as commodities, manufacturers of the equipment face pressure to reduce prices in order to remain competitive. In addition, technological improvements have increased the power of both hardware and software. The combined result of these trends is that GISs are becoming more affordable and, thus, more prevalent. Further, as GISs become more widely used, more people and interest groups will harness this technology to help them build and support their arguments on land-use issues. Because of the ability of GISs to make more information available more quickly, it logically follows that groups that adopt GISs will be able to develop more, and perhaps stronger, arguments to support their positions.

The likely result will be more conflict, not less, as more groups harness GISs to help them build support for their positions on particular land-use issues. In addition, that increased conflict is likely to be cognitive, rather than interest, conflict. Interest conflict should remain steady. Figure 10.2 illustrates the hypothesized change in conflict over time with the introduction of GISs.

A logical question would be to ask how conflict is expected to increase owing to enhanced access to information. After all, if both parties are presented with the same objective data, should not such information resolve misunderstanding and lead to mutually acceptable resolution? If all conflict were simply cognitive (i.e., based on factual

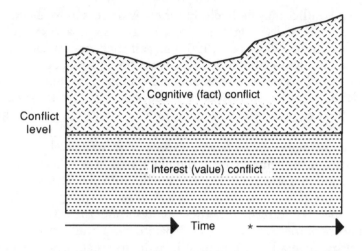

FIGURE 10.2. Hypothetical conflict level after introduction of GIS. Asterisk indicates time of GIS introduction.

disagreement) the answer may perhaps be "Yes." However, value conflict leads to a level of emotion that is not easily influenced simply by objective data. Further, we must realize that conflicting parties often interpret the same data in very different ways. An example would be the different "spins" put on the results of research into the dangers of cigarette smoking when interpreted by the Tobacco Institute versus the Surgeon General's office. All forms of data are only as useful as their interpretation. It is this difficulty with objective interpretation, however, that lies at the heart of conflicts resulting from increased access to information.

GOOD NEWS FOR A DEMOCRACY

While the first response to the hypothesis that widespread introduction of GISs is likely to increase conflict among groups may be negative, we should recognize that in democratic societies conflict is not necessarily bad. In the first place, open disagreement provides evidence that democracy is truly working, that the many voices of the people are heard. When many voices are heard, working out a resolution that is acceptable to all is at least possible, although still often difficult to achieve. At the very least, each group will have had an opportunity to air its concerns. When groups do not have a voice, democracy is not working.

The conflict may be lively, but there is no reason to assume that it will become a disintegrating force. In fact, some authors (North, Koch, & Zinnes, 1960; Pondy, 1967) suggest that conflict can serve an important integrative function. This function comes about as a result of allowing all voices to be heard. In a democracy, the judicial system provides a forum for raising issues. So does voting. Day-to-day arenas such as the free press and mass media provide additional opportunities for interest groups to speak their minds.

Because democracies have a variety of arenas in which interest groups can raise their concerns, the likelihood of conflicts becoming massive and negative is relatively small. Still, from time to time, violent protests, such as those in Los Angeles in spring of 1992, will erupt. However, historically, these protests are the result of long-brewing inequities and conflicts, separate from the introduction of technology. Overall, the potential benefits to a democracy outweigh any foreseeable disadvantages.

CONCLUSIONS

Although it has been suggested that GISs will help to ease conflict among groups by providing more and "better" information, the likelihood of this outcome is difficult to support. On the contrary, as GISs proliferate and "trickle down" to more groups, conflict is likely to grow. This will occur largely as the result of GISs' inability to affect interest (value) conflict at all. Equally important is GISs' potential to increase cognitive (fact-based) conflict by making available more data, which can be rearranged in a variety of ways to create new information. This new information can, in turn, be used to support arguments, and thus will tend to increase conflicts.

The case of the proposed Bailly nuclear generating facility demonstrates how fact-based arguments (cognitive conflict) may be used to promote a particular interest. As we have seen, the judicial system provides an important forum in which cognitive conflict often serves as a surrogate for interest conflict. Any technology that makes available more information that can be used to support cognitive conflict is a valuable tool for a group, such as the Save the Dunes Council, that seeks to derail an unwanted land use and has the knowledge and wealth to mount a legal campaign. For the Save the Dunes Council, as for other groups, a delay is often the next best thing to a clear-cut victory.

Little has been done to answer our questions about the role of GISs in conflict resolution. It is clear, however, that the proliferation of GISs as a means of exploring land-use issues promises to raise this issue as a matter of practical interest. This chapter has presented an initial discussion of the issue.

ELEVEN

❖

GISs in a Democratic Society: Opportunities and Problems

Geographic information systems in a democracy present both opportunities and problems to citizens. As GISs become increasingly powerful, user friendly, and affordable, they become more readily available to a wider range of individuals and interest groups than ever before. Such increased availability of GIS technology is of value in a democracy because it holds the promise of including more voices in important policy debates at all levels of government. As with any technology, problems (technical, social, and legal) are bound to arise as the technology comes into wider use. Initially, the concentration of knowledge about GISs in the hands of technical experts, rather than policy experts, is cause for concern. Ultimately, ensuring that policy experts gain knowledge about the capabilities and limitations of GISs is equally important. In this chapter we will take a theoretical perspective, drawing from literature on organizations regarding the causes, ramifications, and prospects for GISs in a democratic society. In addressing these ramifications, we will also examine some of the more pressing legal concerns that have been generated as a result of the proliferation of GIS data within local, state, and federal governments and their agencies.

Advocates of GISs have seen their efforts reap rewards in recent years as GISs have become increasingly powerful, user friendly, and affordable. Advances in GISs have made the technology more readily available to a wide range of people and organizations who employ them in increasingly diverse applications, from transportation network development and land-use planning to human services delivery. The proliferation of GISs has special significance in a democracy, holding not only the promise of including more voices in important policy decisions but also the threat of concentrating crucial information in the hands of technical experts. The result is a dynamic tension between democracy and technical bureaucracy, a concern raised by Weber in his theory of bureaucracy. While recognizing that trying to predict the future is of questionable value, we also recognize the societal consequences of the implementation of GISs in democratic societies and will attempt to shed light on them.

This chapter begins with a discussion of the democracy–bureaucracy dynamic. We then proceed to develop this theme within the context of what has been called "the Information Age." In the following section we will discuss the opportunities and problems that proliferation of GISs raises for democratic societies. Finally, the chapter concludes with a reminder of the responsibility that developers and advocates of GISs have toward the larger society.

THE DEMOCRACY–BUREAUCRACY DYNAMIC

Concern about the survival of democracy in the face of rapid technical development has long been a topic of interest. Indeed, such popular literature as Orwell's *1984* has raised the specter among the public of technology that is used to threaten the individual civil rights and democratic freedoms of average citizens. Scholarly interest in the survival of democracy has been equally great and has a long history dating back to the work of Hobbes and Locke.

Hobbes and Locke

Writing in the 17th century, Hobbes theorized that "intelligent, selfish human beings prefer to be governed" (quoted in Miller, 1989, 512). His logic rests on the assumption that people prefer to live in an environment where they are protected from the negative aspects of social intercourse, such as violence and robbery. In order to gain this protection, however, people must be willing to submit to the rule of legitimate authority, with consequent limitations on personal freedom. The result, Hobbes posited, is a totalitarian society characterized by "moral reciprocity," wherein its citizens can "begin to realize their potentials for collective morality and prosperity" (quoted in Miller, 1989, 512).

Although he agreed with Hobbes's basic assumptions, Locke challenged the idea that only totalitarian control could guarantee moral reciprocity. Instead, Locke argued, representative democracy would permit citizens to gain the advantages of government without having to submit to the rule of tyrants (in Miller, 1989, 512). Locke's argument could be construed to extend the notion of moral reciprocity to the government itself, positing that if rulers undertake actions that do not have the support of the citizens of the society, then the citizens have the right to rebel. Locke explicitly supported government based on majority rule—that is, democracy.

Democracy can be characterized as being either direct or representative. Direct democracy allows citizens to have a voice in each and every issue that affects them, such as occurs in the New England town-meeting tradition. As another variation of direct democracy, McLean (1989) has included Tiebout's hypothesis (1956) that citizens who are unhappy with local public services will vote with their feet (i.e., move away). Representative democracy, on the other hand, calls for citizens to elect individuals to represent their interests within the government decision-making arena. In this chapter we do not distinguish between direct and representative democracy but do note that representative democracy is the primary operational mode within the United States, which provides the context for this analysis.

The Democracy–Bureaucracy Debate

Regardless of the type of democracy under consideration, the mechanism for implementing the will of the people is not well defined. There is general agreement, however, that the mechanism is an administrative bureaucracy. In recent years, several scholars in public administration (Aberbach & Rockman, 1988; Burke, 1939; Cigler, 1990; Cooper & Lui, 1990; Kearney & Sinha, 1988; Pfiffner, 1987) have elaborated on themes originally developed by Weber (1968) in his theory of bureaucracy. Understanding this debate is useful because of the pervasive role of public bureaucracies and technical experts in the implementation of GISs.

Weber identified two basic principles of administrative bureaucracy that are at odds with democracy. First, bureaucracy provides that only experts can hold a professional position. Democracy, on the other hand, seeks to prevent the development of a closed-status group of officials in the interest of universal accessibility to public office.

Second, bureaucratic theory calls for the bureaucrat to retain his or her position indefinitely. By contrast, democracy seeks to minimize the authority of officialdom in the interest of expanding the "sphere of public opinion as far as practicable" (Weber, 1946, 226). One way of doing this is to require regular and frequent elections of officials.

Over time, Weber posited, the bureaucracy may come to occupy a position of considerable power in society. The power of a bureaucracy derives from three major sources: the expertise of the bureaucrat, the professionalism of the bureaucrat, and the bureaucracy's relationship with its clientele (see Figure 11.1). The first two sources of bureaucratic power are clearly at odds with democracy and lean toward technocracy. The third, however, calls to mind direct democracy.

In the case of the public bureaucracy in a democratic society, the legislature or some other body of elected officials is—nominally at least—the 'master' of the bureaucracy. Yet the "master" rarely has the degree of technical expertise possessed by its "servant," bureaucracy. In practice, this means that the political "master" frequently plays the

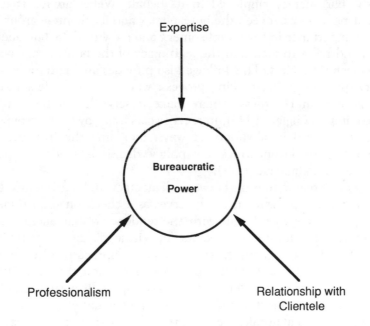

FIGURE 11.1. Sources of bureaucratic power.

role of "dilettante" to the bureaucracy's "expert," a case of the tail wagging the dog. The use of high technology (including GIS) increases the likelihood of this scenario, causing policy makers to depend heavily on experts who operate and control the technology used in policy development and implementation. Such control will certainly influence the policy process.

The second source of the bureaucracy's power is the professionalism of bureaucrats and their ability to guard their knowledge from outsiders. Each profession has a body of knowledge that is uniquely its own. In order to become a member of the profession, the hopeful candidate must go through a well-defined right of passage that ordinarily culminates in some sort of certification process. Technical professions are often very successful, in part because they have learned to guard carefully a relatively small, but highly complex, body of knowledge that is theirs. While limiting entry into the field, certification also helps to preserve the technical credibility of the group within its field of expertise.

The relationship between the bureaucracy and its client group is the third source of bureaucratic power. This relationship is one of mutual dependency wherein the clientele benefits from the expertise of the bureaucracy employed in its behalf. Weber argued that the interdependency between the bureaucracy and its clientele group is of major importance in the establishment and growth of the bureaucratic form and helps to maintain the legitimacy of the bureaucracy within a democratic society. This linkage also provides an auxiliary point of entry into the policy-making process: whoever can understand the bureaucracy and harness it may make it serve his or her interest. According to Gigler (1990), in recent years input by client groups has been formalized in a variety of ways, including the institution of mandatory public hearings on policy decisions made by public organizations (bureaucracies).

The dynamic tension between democracy and bureaucracy gives rise to a situation wherein each serves as a check on the other. For example, Weber found fault with the tendency of bureaucracy as an institution to elevate bureaucrats, by virtue of their superior and specialized knowledge, to the status of a mandarin caste. On the other hand, he saw in the bureaucracy a means to limit the potential for self-interested actions by elected officials (especially corrupt officials). Bureaucrats are subject to the professional judgment of their peers and to their need to maintain the support of clients; therefore, bureaucrats can be held accountable for their actions, just as elected officials are held accountable through the voting process.

Interestingly, Habermas (1970) discussed a technocratic model of organizations wherein politicians become "the mere agent of a scientific intelligentsia" (63). Because of the overwhelming expertise of scientists, Habermas suggested, the ability of politicians to make decisions is illusory at best. Furthermore, experts, advantaged by their superior knowledge, are in a position to limit—either by design or by accidental omission—the knowledge that they pass on to policy makers. This control over knowledge helps technocrats play a crucial role in policy decisions. The politician is further isolated from the decision-making process if the expert is also cast as an advisor. Downs's (1967a) examination of the payoffs of urban data systems supports Habermas's position.

The basic tension within this debate involves the relative roles played by elected officials who make public policy and technical experts who implement (and in some cases make or modify) policy. While elected officials are charged with responsibility for translating the will of the people into public policy, technical experts within government bureaucracies are charged with carrying out the elected officials' (and the voters') mandates. Historically, these two groups have maintained a tension within democratic societies, serving as checks and balances on each other. The infusion of new technologies such as GISs into this picture threatens to upset the balance in favor of the technical experts.

DEMOCRACY AND BUREAUCRACY IN THE INFORMATION SOCIETY

The last 25 years have brought about remarkable technological advances that have, in turn, affected our lives in a variety of ways. Among the advances, those related to computing and communication have come to define what Cleveland (1985) called the "Global Information Society." "The information era," Cleveland asserted, "features a sudden increase in humanity's power to think and therefore to organize" (185). This reorganization from a society based on industrial production to one based on information requires a shift in our underlying assumptions in order to adjust to the distinctive characteristics of information. As Cleveland noted, information is not subject to the laws of thermodynamics (187).

NEW ASSUMPTIONS FOR A NEW ERA

Cleveland identified six significant characteristics of information that will challenge our current ideas about political and administrative arrangements in a democratic society (see Table 11.1). First, information is expandable—so much so that time is the limiting factor in the ability of the human mind to make use of information. Second, information is not resource hungry—it requires relatively small inputs of energy or physical and biological resources, compared to industrial production. Third, information is substitutable—capable of replacing capital, labor, and physical materials. Fourth, information is transportable—in some cases, at or near the speed of light. Fifth, information is diffusive—it tends to leak. Finally, information is shareable—when someone shares information with another, both have the information (Cleveland, 1985, 186–187).

INFORMATION TECHNOLOGY: PANACEA OR PANDORA'S BOX?

Of these characteristics, the "diffusability" and "shareability" of information are most relevant to the proliferation of GISs in a democratic society and fuel what Cleveland referred to as "the erosion of hierarchies" (187). The current political/administrative hierarchy is based on the characteristics of physical resources, which are subject to laws of thermodynamics and economic valuing based on scarcity, and

TABLE 11.1. Characteristics of Information

1. *Information is expandable*. Information expands as it is used.

2. *Information is not resource hungry*. The production and distribution of information are generally energy efficient.

3. *Information is substitutable*. Through robotic and other forms of automation, information can replace more traditional sources of capital and labor.

4. *Information is transportable*. The power of electronic transfer has rendered the notion of "remoteness" essentially obsolete.

5. *Information is diffusive*. Information leaks in a continuous and pervasive manner and is self-generating. The more it leaks, the more we tend to have.

6. *Information is shareable*. To be information rich means to share information. When information is sold or shared both parties now possess it.

Note. Adapted from Cleveland (1985).

which, therefore, are easily controlled. The erosion of hierarchies suggests that even as we develop standards, conventions, and codes of ethics regarding the use and transfer of information (including computerized information), tight control of information among a ruling elite will not be easily achieved. Based on this line of reasoning, Cleveland projected a rosy future for democracy in the information age, suggesting that average people will use information technology to amplify their voices in matters related to their own governance.

Other scholars have offered different ideas. Barber (1988) painted three different scenarios of life in the information society: Pangloss, Pandora, and Jeffersonian. The Pangloss scenario, named for the irrepressibly optimistic and naive Dr. Pangloss of Voltaire's *Candide*, depicts a utopian society in which information technologies provide an avenue for direct majority control of society through the use of these technologies (180–185). The recent challenge by Cable News Network to established network news in the United States and the "electronic town meetings" proposed by 1992 U.S. Presidential candidate Ross Perot are examples of the promise of Pangloss, which Barber believed is overly naive and optimistic. The Pandora scenario (185–187) envisions the results of a government that deliberately sets out to control information technology with *1984*-like outcomes. In between these is the Jeffersonian scenario (187–190), which views information technology as having the potential to "offer powerful assistance to the life of democracy" (187).

Downs (1967a) suggested that the wide-scale implementation of urban data systems will strengthen the civil servants who control the systems at the expense of elected officials, thus undermining democracy. However, this strengthening will not be universal. Downs noted that "automated systems tend to transmit many data directly from events themselves to top-level officials, or at least to reduce the number of intermediate steps separating these two extremes. As a result, lower- and intermediate-level officials often have little or no opportunity of "filtering" important data before they reach top-level officials" (1967a, 207). Downs has defined filtering as "either distorting information by altering or leaving out parts of it, or adding personal interpretations to it, or both" (207). The implication is that middle-level managers may be squeezed out of the decision-making loop. The recent spate of layoffs among middle-level managers in the United States may lend credence to Downs's prediction. Echoing Downs, O'Dowd (1988) suggested that "those controlling the technology also possess the power to influence the manner in which the information is actually received, thus public opinion may be shaped and manipulated, and social control exerted" (10). This is consistent with Barber's Pandora scenario but overlooks Cleveland's observation about the leakage of information.

Pointing specifically to the problem of unemployment created by the proliferation of computers in society, Street (1988) speculated on additional negative side effects of the rise of the information technology (8). We are all familiar with the role that computerization plays in de-skilling and unemployment as computers (and robotics) replace human workers. Street linked this problem with another more serious potential problem in a democratic society, disenfranchisement, suggesting that those who are unemployed are no longer active participants in the economic system. This situation, left unattended, Street argued, eventually creates problems of legitimacy for the ruling power (9).

Similarly, Bevan (1988) recognized that the inability of "ordinary people" with a "good understanding of their needs" to verbalize or formalize their needs within a policy context that includes computerization may put them at a disadvantage in the information age (334). Bevan suggested that as information technology becomes more prevalent, data that can most easily be accommodated to the technology itself may take priority over information for ordinary people. The obvious result would be increased control of policy by technocrats with a compensating loss of democratic control. Metzendorf (1988) recommended a strategy to address the problems of illiteracy and computer inequity among the poor as a means of facilitating democracy.

EVIDENCE FROM REAL LIFE

What does this range of scenarios suggest? Real-life examples provide some clues about what awaits us as the information age matures, generally, and as GISs become more commonplace, specifically. Within the computer software industry, we can point to the difficulty of combating the all-too-common practice of pirating software as an example of what Cleveland described as the leakage and shareability of information. Pirating software increases the computing capabilities of those who steal it, and yet the software developer still has the software and can continue to sell it to others. Contrast this with the loss sustained by the automobile dealer when a thief steals a car from the dealership lot; obviously, the missing (stolen) car cannot be sold. It seems likely that average computer users will continue to share and exchange pilfered software, as well as legitimate "shareware" and databases, putting more information and greater ability to examine it in the hands of increasing numbers of people. While this may be bad news from a capitalistic perspective, it is good news from the democratic perspective.

In addition, fully integrated databases like the "TIGER Files" of the United States Bureau of the Census hold the promise of reliable data generally available at a reasonable price. Again, the value of the shareability of data comes to life. Already becoming available is shareware that will make these data accessible to a wide range of groups and individuals employing a vast array of hardware and software configurations to achieve diverse goals and objectives. That these data are being made available to the general public is in and of itself a good omen for democracy; the role that the data may play in future policy decisions will tell the full story.

It is important to note that the Freedom of Information Act in the United States looms as an important administrative tool in the ability of Bevan's "ordinary people" with a "good understanding of their needs" to harness GISs in pursuit of specific policy objectives. This act makes it possible to obtain pubic information that might otherwise be beyond reach, thus contributing to the society Orwell envisioned in 1984. However, the absence of a Freedom of Information Act should not be viewed as an impenetrable barrier to information exchange. This issue serves as one example of the legal considerations that will be explored later in this chapter.

A remarkable example of the transportability, diffusability, and shareability of information in the absence of a Freedom of Information Act occurred in conjunction with the failed Chinese student rebellion in Tienanmen Square in June 1989. It has been widely reported that fax machines played a key role in gaining international support for the students' efforts as well as in keeping the world informed about events during those electrifying days. Similarly creative means for sharing computerized information may overcome governmental holds on information. When one considers the amount of data that a single floppy disk for a personal computer holds, the potential for sharing (even smuggling) information becomes readily apparent.

Similarly, Wilke (1993) reports in the *Wall Street Journal* that the day-to-day use of electronic mail and other "groupware" in a growing number and variety of venues in the United States (and elsewhere) is

helping to "erode [the] hierarchical nature of workplace culture" (A1).
Wilke notes that

> because they enable hundreds of workers to share information
> simultaneously, groupware networks can give lowly office workers
> intelligence previously available only to their bosses. Networks also can
> give the rank-and-file new access: the ability to join in on-line
> discussions with senior executives. In these interactions, people are
> judged more by what they say than by their rank on the corporate ladder.
> (A1)

Wilke cautioned, however, that the proliferation of networks can
also have negative consequences. For example, some members of the
network may not share their best ideas with other members of the
network, and instead hoard them as a way to gain or maintain an
advantage in the organization. In some instances, networks may result
in information overload, as users flood the system with information of
variable importance. Networks may also be used as a tool of management to monitor the activities of employees (Wilke, 1993, A7).

The implication of Wilke's case-study evidence is that the adoption of information technology may be either good or bad for democracy. "This new electronic landscape can foster an egalitarian sense of
empowerment among employees. Or it can be a tool of authoritarian
managers, leading to loss of workplace privacy. For better or for worse,"
he noted, quoting Bruce Hasenyager, senior vice president of Chemical
Banking Corporation, " 'it's a powerful means of amplifying the style
and character of a company and its managers' " (B7).

And yet we also know that access to computers is far from
universal. In fact, full reading literacy—let alone computer literacy—
remains an aspiration for many nations, including the United States.
Differential funding levels for education result in some school districts
in which basic and computer literacy are nearly universal and others in
which the drop-out rate is high and where even those who graduate
from high school may have reading and analytical skills far below the
level expected for persons of their age. The ability of individuals in such
environments to participate in the democratic process may be seriously
impaired; at the very least, we should not be surprised if they join the
ranks of those apathetic toward or alienated from the democratic
process.

Evidence so far indicates that the erosion of hierarchies suggested
by Cleveland is indeed occurring in this information age, boding well

for democracy. Control of information and information technology by the ruling elite seems unlikely at this stage in existing democracies, although the same may not be the case in countries currently living under totalitarian regimes. At the same time, we must not be over optimistic about the power of information technology to bring disen-franchised groups into the democratic process. And unfortunately, we should recognize that the same people who have historically been denied access to traditional tools of democracy continue to be frozen out of the information society.

THE PROLIFERATION OF GISs IN A DEMOCRACY

We have noted that GISs are already playing a growing role in public policy decisions in a variety of geographic and functional areas. Public organizations are implementing GISs at an increasing rate with a clear potential to enhance or impede democratic processes. As GISs become more affordable, more user friendly, and more powerful, individuals and groups may employ them to perform analyses in support of specific policy objectives, thus enhancing democracy. Making this scenario a reality, however, will not happen overnight but will take time and effort. Furthermore, it will require balancing two areas consistent with the democracy–bureaucracy dynamic: substantive issues and technical (computer) expertise.

In the exposition of the democracy–bureaucracy dynamic, we discussed bureaucratic expertise and professionalism as impediments to democracy. The technical expertise of bureaucrats has been known to force legitimately elected leaders to defer to the judgment of technical experts, thereby limiting policy choices available to them and, by extension, to their constituents (Yellin, 1981). When technical expertise is combined with control of a tool used in the policy-making process (such as a GIS), the power of the bureaucracy is increased to the detriment of democracy.

In the case of GISs, the fact that the technology is computer-based has heretofore limited access to GISs to the computer literate. Moreover, it has given rise to a group of "experts" who know all the commands required to operate a specific GIS but who have difficulty conceptualizing an approach to real-life problems using a GIS. This control by technical experts is consistent with the Weberian theory of bureaucracy and represents a real threat to democracy.

Two trends offer a potential remedy to the problem of isolated technical expertise in GISs. First, there has been an increase in the overall rate of computer literacy and access among ordinary people. This trend offers the promise that people with substantive knowledge of the issues will take the bull by the horns and increase their technical expertise in GISs. Second, computers themselves (and GISs) have been made increasingly easy to use and operate. This makes GIS technology increasingly accessible to people with substantive knowledge of issues. These trends may be products of the information age as described by Cleveland. Whatever their origin, they offer the prospect of deemphasizing the role of technical expertise and returning the balance between issues and expertise—and, by extension, between democracy and bureaucracy.

It appears that the field of GISs is currently in a transitional stage, moving from an emphasis on technical expertise to substantive issue knowledge. Increasingly, we are hearing of employers of GIS professionals (both vendors and organizations using GIS) seeking candidates for employment who have substantive knowledge of particular issue areas and of spatial analysis. The vendors recognize that it is easier to transfer technical expertise about GISs than to educate technical experts in substantive issue areas in which GISs are being employed. This increasing emphasis on substantive issues among GIS professionals must be perceived as a trend that supports democracy.

ETHICAL IMPLICATIONS OF THE PROLIFERATION OF GIS

Geographers in particular have begun to discuss the ethical implications of the proliferation of GISs. In late 1993, the National Center for Geographic Information and Analysis sponsored a workshop called "Geographic Information and Society," the first such forum devoted specifically to this important topic. As a follow-up, conference organizer Tom Poiker established an electronic mail bulletin board (GISOC list on Internet) to foster ongoing communication about this topic. At the same time, the *Professional Geographer* published an "Open Forum: Automated Cartography in 1993" (Lord, 1993), which covered similar ground as the workshop.

The ethical implications of the widespread adoption of GISs are potentially vast, with significant moral implications. Immediately we are reminded of Neil Smith's (1992) concern about the use of the GIS as a cog in the war machine (covered in more detail in Chapter Nine). In addition, Pickles (1993) has raised concerns about ethical issues arising from the use of GIS in social surveillance and marketing by the state, the military, and private business (453).

As Goodchild noted, "we cannot escape the fact that GIS has had a significant impact on many kinds of human activity, from local government to the management of utilities, and has changed the role and nature of geographic information profoundly" (1993, 445). Sheppard (1993) and Pickles (1993) in particular have raised concerns about the failure of the GIS community specifically and geographers at large to question the blind assumption of technological neutrality that so often accompanies GIS implementation. In fact, as noted in the preface to this book, much of the existing case-study literature on GISs totally ignores the institutional and societal implications of the technology.

If GISs behave as Wilke (1993) found networks to behave—that is, if GISs amplify the underlying culture (whether democratic or authoritarian) of the institution or society that implements it—then we will not be surprised at the widely divergent views on the ethical ramifications of GIS implementation. Instead, we will recognize that GIS implementation is a value-laden endeavor and act accordingly.

LEGAL RAMIFICATIONS OF GISs

With the proliferation of GISs within many public organizations and agencies, a great deal of previously inaccessible data has, within the past 10 years, become increasingly available. The enhanced sophistication in GIS technology and wider accessibility to users has led, however, to a number of institutional and legal issues that have rather quickly become quite relevant (Archer & Croswell, 1989). In this problem, the concerns of GIS-using organizations has mirrored recent difficulties in the medical field—that is, technological innovation that proceeds at a rapid pace often has the effect of creating a plethora of new legal and ethical ramifications. For example, public access to GIS data offers the fuel for a serious debate on the relative merits of right to privacy versus freedom of information. In this section we will explore some of the basic legal concerns about the use of GIS data by organizations and private citizens. Specifically, we hope to point to some of the important questions in the legal debate on the use of geographic information within a democratic society.

Speaking in a general sense, the central legal problems can be summarized in the following four questions: (1) From a legal standpoint, who is the owner of a database? Can database material be copyrighted? (2) Under the U.S. Freedom of Information Act, what are the rights of private citizens to obtain state and federal governmental databases? (3) How can disclosure laws be reconciled with the right to privacy? (4) What does the law suggest concerning the liability of those using, sharing, or distributing data? Each of these questions has important ramifications for the long-term acceptance and continued development of geographic information technologies. Because they are so central to the future of GISs, these questions deserve to be explored in some detail. However, it is not our purpose in this section to offer definitive findings on each of the above questions. Many of these issues will continue to be argued in the courts for years to come. If, however, we are able to gain a better understanding of the *nature* of the questions themselves, it will allow us to draw our own informed conclusions regarding the role of GISs in a democracy.

DATABASE COPYRIGHT

The problem of database copyright represents a significantly different issue than that of copyrighting basic computer software. In the latter case, there is general agreement that the author of the work is considered the owner of the copyright. Further, if the author is employed by a company and creates the software within the scope of employment, the software's owner is the employing company. The issue is not nearly as straightforward in the area of database copyright, however. Past legal precedent suggests that gathered "facts" are not subject to copyright protection, implying that the act of gathering data into a centrally compiled database may not be copyrighted. On the other hand, if the organization or individual can demonstrate that the acquired data has been selected, coordinated, or arranged in a unique manner, the information may be copyrighted. The problem here is obvious. Because there has been no definitive ruling on the ownership of databases, it is difficult to protect or claim ownership to database material, particularly for an organization attempting to prove that its staff has taken accessible data and "arranged" it in a manner that constitutes copyright protection. The problem is compounded by the type of agency that controls the database. For example, for a database existing in the public domain, no permission is required to use the work of the U.S. government. On the other hand, the work of foreign, state, or local governments may be subject to copyright. As a result, the equation for use of public databases has become extremely complex not only because of the difficulty in proving that the database material is truly unique—and thus merits copyright protection—but also because of the varying rules for use of databases by differing government agencies.

THE U.S. FREEDOM OF INFORMATION ACT

The Freedom of Information Act (1982) applies, in part, to the disclosure of "Public Information; agency rules, opinions, orders, *records*, and proceedings." The essence of the act is that the federal government must make available to the public information that falls within the broad context of material that each agency has gathered, developed, or stored (e.g., databases). While the law itself contains some degree of ambiguity in language and interpretation, subsequent case law has refined the original intent of the act. In effect, case law states that any form of information that can be defined as a "record" must be disclosed—that is, made available to the public at fair price. The only exceptions are those records that fall within one of nine narrowly defined categories—for example, trade secrets or information exempted from disclosure by statute.

As a result of the case law interpretations of the Freedom of Information Act, databases have been classified as records and, as such, are subject to disclosure. The practical implications of this act are that a number of federal databases (e.g., census data) are readily obtainable from various governmental agencies for a fee. Subject to a fair time period to allow for search and review and duplication, most federally generated information can be accessed. One of the areas of continued debate resulting from attempts to obtain federal databases has been the interpretation of the concept of "fair price." In recent years, some private organizations have charged that federal agencies, in an attempt to retain control of their data, have begun escalating prices beyond the range of what would be considered "fair." This controversy has threatened to lead to a new round of litigation as advocates for information disclosure charge the government with willful intransigence.

RIGHTS OF PRIVACY AND CONFIDENTIALITY

One of the continuing controversies resulting from the proliferation of publicly accessible databases has to do with the rights of the individual to privacy (the "right to be let alone") with the rights of the citizenry to accessible information. Geographic information, as contained in a number of databases, has the potential to contain a great deal of personal information. For example, census files that contain information about personal income are readily obtainable by any private telemarketing or direct mail firm willing to pay for the data. The original Freedom of Information Act contained provisions at a number of points concerning the rights of privacy. In fact, one of the nine exceptions to the act included limiting access to records that "could reasonably be expected to constitute an unwarranted invasion of personal privacy" (Freedom of Information Act, 105). Common law has long upheld the tenet that the government shall not interfere in intimate personal activities nor allow such intrusion by other private citizens. Unfortunately, most of the established code of law on privacy has concerned itself with conflicts among individuals. Conflicts of privacy in the computer age affect everyone.

In an effort to better safeguard the privacy of individuals from governmental agencies and private organizations, a number of pieces of legislation have been passed since the 1970s. While state and local governments have enacted similar laws, the more well known statutes are federal. These acts include the Fair Credit Reporting Act (1970), Family Education Rights and Privacy Act (1974), Federal Privacy Act (1974), Tax Reform Act (1976), Right to Financial Privacy Act (1978), Privacy Protection Act (1978), Electronic Fund Transfer Act (1980), and the Electronic Communications Privacy Act (1986). With some exceptions, one of the interesting aspects of the above list of privacy protection legislation is that much of it is the direct result of advances in the areas of information technology and computerization of databases. It could be argued with some justification that the problem with the enacted laws to date is that they do not go far enough in protecting the individual from unwanted intrusions. While they limit the activities of government agencies, they do not affect, to any great degree, the operations of firms holding privately owned databases.

The ability of private organizations to collect, develop, and use databases constitutes one of the greatest threats to individual privacy today. Because the law regarding the right of private companies to access this data is often vague or underdeveloped, there is a wide range of interpretation for privacy standards at local, state, and federal levels. Consider the case, for example, of a real estate broker or direct marketing firm requesting from a local government a copy of its census database for the purpose of accessing and contacting all residents who make an income in excess of $100,000. While many people would interpret a government providing this information to be a breach of privacy, the law may not support this contention. In fact, the law as it is now interpreted may actually require the local government to make this information available at a fair cost.

There are other more insidious threats to privacy as well. For example, Shapiro (1993) reported in the *Wall Street Journal* that "Philip Morris Company's huge database [of 26 million smokers] is playing an increasingly important role in the cigarette giant's marketing efforts, raising concerns among privacy experts." Part of the concern lies in the relative freedom of this private corporation to handle its database as it wishes, whether holding it close or selling it to others. The Philip Morris database includes both home addresses and "detailed personal information" (B1). The smokers themselves provided much of this information in order to qualify for gifts from the Philip Morris Company. According to Shapiro, the United Smokers Association (an organization promoting smokers' rights) has two major concerns about repercussions against people whose names are included in the database, if the database were to fall into the wrong hands: (1) some employers do not hire smokers and (2) some insurance companies charge higher rates to smokers (B6). Shapiro reported that a spokeswoman from Philip Morris "rejects such security concerns and says the list is a closely guarded secret" (B6). Even so, Shapiro noted that the coupons and other "benefits" that Philip Morris mails to the homes of people whose names appear on its database may thwart efforts to quit smoking. Even in the best-case scenario, the very existence of this database has moral implications, since the purpose of it is to promote a known hazardous human activity. No doubt Pickles (1993) would find the Philip Morris Company's database of 26 million smokers troubling from an ethical standpoint.

What, if anything, can be done about potential abuses of privacy through the widespread availability of database information? The Congress is attempting to head off continued abuses of individual privacy by private firms. A bill is currently before the House of Representatives (HR 685) that would create a data-protection board to keep personal information about consumers from telemarketing and direct mailing firms. As has recently been noted, the controversy regarding the drafting of privacy protection legislation lies in attempting to draw a line between the permissible exchange of facts in support of commercial goals and impermissible intrusion (Onsrud, 1992; Epstein & McLaughlin, 1990; Trubow, 1991). Until this distinction is effectively drawn, the resolution of privacy issues poses one of the greatest challenges and potential threats to the continued development and widespread use of geographic information among GIS professionals today.

LIABILITY AND THE USE, SHARING, AND DISTRIBUTION OF DATA

The final legal aspect in the use of geographic information to be considered involves the potential liability that can accrue to those owning or making use of databases. Let us assume, for example, that a city government in Florida develops a database of zoning districts for commercial building and sells that database to a local developer. Let us further suppose that some of the information in the database concerning property-line measurements was inaccurate and, because the builder was unaware of this inaccuracy, allowed the database information to serve as the basis for developing building plots. As the measurement errors are discovered and the builder takes on substantial rebuilding costs to conform to true plot surveys, the question of who is liable for this extra expense comes into focus. Is the owner of the database responsible for any errors in the data or does liability accrue to the individual or organization that originally collected and coded the data? Or does the idea of caveat emptor apply, whereby users of geographic data must assume responsibility for possible errors in the data set? The whole area of informational liability is one that has evolved into a major legal controversy.

In the past, municipalities have never faced such a high level of liability exposure. Certainly, they have always been held accountable for maintaining their public responsibilities (e.g., infrastructure repair and protecting the public good) as well as for any malfeasance on the part of their employees. However, as a result of the expanded use of geographic information, a new area of legal liability has opened up. The reason is that many local and county governments and agencies are looking to sell geographic data as a potential source of revenue. Because they are using geographic information as a profit-making device, the courts have generally held these municipalities to a higher level of liability. In effect, GISs have opened a potential can of worms by making this data so available and readily marketable.

The current state of liability law, as it applies to geographic information, is quite complex. The difficulty that the courts are finding is in attempting to show exclusive culpability (Onsrud & Hintz, 1991). In other words, exactly where does the responsibility lie when errors are discovered in database information? With most database errors, there are three obvious candidates for liability: the database manager, those delivering data to the system, and the users of the system. Unfortunately, the problem of establishing culpability is compounded by the nature of databases themselves. Because databases are usually interactive, it is relatively easy to alter data after the fact to "show" that there was no error in the original data. The obvious assumption then is that any error that occurred happened as a result of operationalizing the information. In other words, the argument is often heard: "Don't blame the database. Blame the person using the information." Until systems are in place to allow experts to track the sources of errors, liability law will continue to affect the operations of municipalities in their commercial ventures with geographic information databases.

CONCLUSIONS: GISs AND SOCIETAL RESPONSIBILITY

Geographic information systems can and will play an important and positive role in democratic societies. The fact that this technology is explicitly employed in the development and implementation of public policy makes its connection with democracy clear. The use of GISs in political reapportionment and redistricting provides unmistakable evidence of its value in a democratic society. GIS professionals must first recognize that their work has implications for democracy and must accept the societal responsibility that their work entails.

Understanding the dynamic between democracy and technical bureaucracy should increase our awareness of the need to balance technical expertise with substantive issue knowledge. GIS professionals whose contributions have made GIS more user friendly have played an important role in moving the technology toward greater democracy. Those who implement GIS with the user in mind and remind GIS professionals of their responsibility to the user also help to keep those professionals honest.

Educators who emphasize the importance of substantive issue knowledge and spatial analysis as essential underpinnings of GISs help students to develop an appropriate sense of balance between knowledge of GISs and an understanding of substantive issues. Students who develop this sense of balance are less likely to allow the tail to wag the dog and are more likely to be aware of the implications of GISs as tools in the resolution of important policy issues in democratic societies.

However, we must also play an active role in realizing the full potential of GISs in a democratic society. Geographic information systems are not implemented in a vacuum: they are implemented within the context of a political structure that seeks to meet the needs of members of society at large. GIS professionals might consider making available their technology and substantive knowledge in service to groups and individuals who previously have had little voice in the democratic process, helping them to make their voices heard. GIS professionals might explore ways to share their knowledge with the wider public. At the very least, GIS professionals must acknowledge and accept their social responsibility. If they do so, then the future for GISs in a democratic society is bright.

❖
References

Aangeenbrug, R. T. 1991. A critique of GIS. In D. J. Maguire, M. F. Goodchild, & D. W. Rhind (Eds.), *Geographical Information Systems: Principles and Applications* (Vol. 1) (New York: Wiley), 101–107.

Aaron, H., & McGuire, M. 1970. Public goods and income distribution. *Econometrica, 36*, 907–920.

Aberbach, J. D., & Rockman, B. A. 1988. Mandates or mandarins? Control and discretion in the modern administrative state. *Public Administration Review, 48*, 606–612.

Abler, R. F. 1987. What shall we say? To whom shall we speak? *Annals of the Association of American Geographers, 77*, 511–524.

American Planning Association. 1991, June 15. *Jobmart* (Chicago: Author).

Anderson, J. C., & Narus, J. A. 1986. Toward a better understanding of distribution channel working relationships. In K. Backhaus & D. Wilson (Eds.), *Industrial Marketing: A German–American Perspective* (Berlin: Springer-Verlag), 320–336.

Antenucci, J. C., Brown, K., Croswell, P. L., & Kevany, M. 1991. *Geographic Information Systems: A Guide to the Technology* (New York: Van Nostrand Reinhold).

Archer, H., & Croswell, P. L. 1989. Public access to geographic information systems: An emerging legal issue. *Photogrammetric Engineering and Remote Sensing, 55*, 1575–1581.

Argote, L. 1982. Input uncertainty and organizational coordination in hospital emergency units. *Administrative Science Quarterly, 27*, 420–434.

Arnoff, E. L. 1971. Successful models I have known. *Decision Sciences, 2*, 141–148.

Aronoff, S. 1989. *Geographic Information Systems: A Management Perspective* (Ottawa, Ontario: WDL Publications).

Association of American Geographers. 1991. *AAG Annual Meeting Program* (Washington, DC: Author).

Baker, B. N., Murphy, D. C., & Fisher, D. 1983. Factors affecting project success. In D. I. Cleland & W. R. King (Eds.), *Project Management Handbook* (New York: Van Nostrand Reinhold), 902–919.

Barber, B. R. 1988. Pangloss, Pandora or Jefferson? Three scenarios for the future of technology and democracy. In R. Plant, F. Gregory, & A. Brier (Eds.), *Information Technology: The Public Issues* (London: Manchester University Press), 177–191.

Bean, A. S., Neal, R. D., Radnor, M., & Tansik, D. A. 1975. Structural and behavioral correlates of implementation in U.S. business organizations. In R. L. Schultz & D. P. Slevin (Eds.), *Implementing Operations Research and Management Science* (New York: Elsevier), 77–132.

Benbasat, I., Goldstein, D. K., & Mead, M. 1987. The case research strategy in studies of information systems. *MIS Quarterly, 11*(3), 369–386.

Bennett, R. J. 1980. *The Geography of Public Finance* (London: Methuen).

Berlant, J. L. 1975. *Profession and Monopoly: A Study of Medicine in the United States and Great Britain* (Berkeley: University of California Press).

Bevan, E. B. 1989. The task for the new professionalism. In B. Glastonbury, W. LaMendola, & S. Toole (Eds.), *Information Technology and the Human Services* (New York: Wiley), 332–341.

Bie, S. 1984. Organizational needs for technological advancement. *Cartographica, 21*, 44–50.

Bonoma, T. V. 1985. Case research in marketing: Opportunities, problems, and a process. *Journal of Marketing Research, 22*, 199–208.

Brennan, G. 1976. The distributional implications of public goods. *Econometrica, 44*, 391–399.

Brill, E. D., Flach, J. M., Hopkins, L. D., & Ranjithan S. 1990. MGA: A decision support system for complex, incompletely defined problems. *IEEE Transactions on Systems, Man and Cybernetics, 20*, 745–757).

Buntz, C. G., & Radin, B. A. 1983. Managing intergovernmental conflict: The case of human services. *Public Administration Review, 43*, 403–412.

Burke, J. P. 1989. Reconciling public administration and democracy: The role of the responsible administrator. *Public Administration Review, 49*, 180–185.

Burroughs, P. A. 1986/1989. *Principles of Geographical Information Systems for Land Resources Assessment* (Oxford, UK: Clarendon Press).

Camillus, J. C. 1986. *Strategic Planning and Management Control* (Lexington, MA: Lexington Books).

Catsambas, T. 1982. Substitutability, separability, and the distributional implications of public goods. *Public Finance Quarterly, 10*, 333–353.

Cayer, N. J., & Weschler, L. F. 1988. *Public Administration: Social Change and Adaptive Management* (New York: St. Martin's Press).

Chrisman, N. R. 1987. Design of geographic information systems based on social and cultural goals. *Photogrammetric Engineering and Remote Sensing, 53*, 1367–1370.

Chrisman, N. R. 1987. Fundamental principles of geographic information systems. *Proceedings of AUTO/CARTO*.

Churchman, C. W., & Schainblatt, H. A. 1965. The researcher and the manager: A dialectic of implementation. *Management Science, 11*, 1369–1387.

Cigler, B. A. 1990. Public administration and the paradox of professionalism. *Public Administration Review, 50*, 637–653.

Clare, D. A., & Sanford, D. G. 1984. Cooperation and conflict between industrial sales and production. *Industrial Marketing Management, 13*, 163–169.

Clark, G. L. 1981. Law, the state, and the spatial integration of the United States. *Environment and Planning—A, 13*(10), 1197–1232.

Cleveland, H. 1985. The twilight of hierarchy: Speculations on the global information society. *Public Administration Review, 45*, 185–195.

Cook, T. D., & Campbell, D. T. 1979. *Quasi-Experimentation: Design and Analysis for Field Settings* (Boston: Houghton Mifflin).

Cooper, T. L., & Lui, T. L. 1990. Democracy and the administrative state: The case of Hong Kong. *Public Administration Review, 50*, 332–344.

Cowen, D. J. 1987. GIS versus CAD versus DBMS: What are the differences? *Proceedings of GIS '87*, Falls Church, VA.

Cromley, R. G. 1983. Automated geography: Some problems and pitfalls. *Professional Geographer, 35*, 340–341.

Croswell, P. L. 1989. Facing reality in GIS implementation: lessons learned and obstacles to be overcome. *Proceedings of the 27th Annual Urban and Regional Information Systems Association, 4*, 15–35.

Cyert, R. M., & March, J. G. 1963. *A Behavioral Theory of the Firm* (Englewood Cliffs, NJ: Prentice-Hall).

Dansby, B. 1991. Recovering GIS development costs by copyright use. *GIS World, 4*(2), 100–101.

Danziger, J. N., Dutton, W. H., Kling, R., & Kraemer, K. L. 1982. *Computers and Politics: High Technology in American Local Governments* (New York: Columbia University Press).

De Blij, H., & Muller, P. O. 1992. *Geography: Regions and Concepts* (6th rev ed.) (New York: Wiley).

Deacon, R., & Shapiro, P. 1975. Private preference for collective goods revealed through voting on referenda. *American Economic Review, 64*, 943–955.

DeLeon, W. H., & McLean, E. R. 1992. Information systems success: The quest for the dependent variable. *Information Systems Research, 3*, 60–95.

Deshpande, R., & Zaltman, G. 1987. A comparison of factors affecting use of marketing information in consumer and industrial firms. *Journal of Marketing Research, 21*, 114–118.

Deutsch, M. 1949. A theory of cooperation and competition. *Human Relations, 2*, 129–152.

Deutsch, M. 1958. Trust and suspicion. *Journal of Conflict Resolution*, 2, 265–279.

Dickinson, H. J., & Calkins, H. W. 1988. The economic evaluation of implementing a GIS. *International Journal of Geographical Information Systems*, 2(4), 307–327.

Dickinson, H. J., & Calkins, H. W. 1990. Comment on "Concerning the economic evaluation of implementing a GIS." *International Journal of Geographical Information Systems*, 4(2), 211–212.

Dobson, J. E. 1983a. Automated geography. *Professional Geographer*, 35, 135–143.

Dobson, J. E. 1983b. Reply to comments on "Automated Geography." *Professional Geographer*, 35, 349–353.

Dobson, J. E. 1993. The geographic revolution: A retrospective on the age of automated geography. *Professional Geographer*, 45(4), 431–439.

Doll, W. J., & Torkzadeh, G. 1988. The measurement of end user computing satisfaction. *MIS Quarterly*, 12, 259–276.

Douglas, M. 1986. *How Institutions Think* (Syracuse, NY: Syracuse University Press).

Downs, A. 1967. A realistic look at the final payoffs from urban data systems. *Public Administration Review*, 27, 204–210.

Downs, A. 1967. *Inside Bureaucracy* (Boston: Little, Brown).

Downs, G. W., & Mohr, L. B. 1976. Conceptual issues in the study of innovations. *Administrative Science Quarterly*, 21, 700–714.

Dumaine, B. 1990, May 7. Who needs a boss? *Fortune*, 52–60.

Dwyer, F. R., & Oh, S. 1987. Output sector munificence effects on the internal political economy of marketing channels. *Journal of Marketing Research*, 24, 347–358.

Dwyer, F. R., Oh, S., & Welsh, M. A. 1985. Environmental relationships of the internal political economy of marketing channels. *Journal of Marketing Research*, 22, 397–414.

Engle, J. R. 1983. *Sacred Sands: The Struggle for Community at the Indiana Dunes* (Middletown, CT: Wesleyan University Press).

Epstein, E. F. 1990. Access to information: Legal issues. *Proceedings of the XIX Congress of the International Federation of Surveyors*, 3, 92–99.

Epstein, E. F., & Duchesneau, T. D. 1984. *The Use and Value of a Geodetic Reference System* (Orono, ME: University of Maine and Federal Geodetic Control Committee).

Epstein, E. F., & McLaughlin, J. D. 1990. A discussion on public information. *ACSM Bulletin*, 128, 33–38.

Faulhaber, G. R. 1975. Cross-subsidization: pricing in public enterprises. *American Economic Review*, 65(5), 966–977.

Fayol, H. 1916/1949. *General and Industrial Administration* (London: Pitman).

Fayol, H. 1929. *General and Industrial Management* (J. A. Conbrough, Trans.) (Geneva: International Management Institute).

Gage, R. W. 1984. Federal regional councils: Networking organizations for policy management in the intergovernmental system. *Public Administration Review, 44*, 134–144.

Galbraith, J. R., & Nathanson, D. A. 1978. *Strategic Implementation: The Role of Structure and Process* (Dallas, TX: Business Publications).

Galletta, D., & Lederer, A. L. 1989. Some cautions on the measurement of user information satisfaction. *Decision Sciences, 20*, 419–438.

Garreau, J. 1981. *The Nine Nations of North America* (Boston: Houghton Mifflin).

George, A., & McKeown, T. 1985. Case studies and theories of organizational decision making. In L. Sproull & P. Larkey (Eds.), *Information Processing in Organizations* (Greenwich, CT: JAI Press), 21–58.

Gersick, C. J. C., & Davis-Sacks, M. L. 1990. Summary: Task forces. In J. R. Hackman (Ed.), *Groups That Work (and Those That Don't)* (San Francisco: Jossey-Bass), 146–154.

Gersmehl, J. 1985. The data, the reader and the innocent bystander: A parable for map users. *Professional Geographer, 37*(3), 329–334.

Gillespie, S. R. 1991. *Measuring the benefits of GIS use.* Paper presented at the ACSM-ASPRS Fall Convention, Atlanta.

Goodchild, M. F. 1988. Stepping over the line: Technological constraints and the new cartography. *American Cartographer, 15*(3), 311–319.

Goodchild, M. F. 1991. The technological setting of GIS. In D. J. Maguire, M. F. Goodchild, & D. W. Rhind (Eds.), *Geographical Information Systems: Principles and Applications* (Vol. 1) (New York: Wiley), 45–54.

Goodchild, M. F. 1993. Ten years ahead: Dobson's automated geography in 1993. *Professional Geographer, 45*(4), 444–446.

Goodchild, M. F., & Getis A. 1991. *Introduction to Spatial Analysis.* Notes accompanying a workshop of the same name held in conjunction with the annual GIS/LIS Conference, Atlanta, GA.

Goodchild, M. F., & Rizzo, B. R. 1987. Performance evaluation and work-load estimation for geographic information systems. *International Journal of Geographic Information Systems, 1*(1), 67–76.

Goodsell, C. T. 1983/1985. *The Case for Bureaucracy: A Public Administration Polemic* (Chatham, NJ: Chatham House).

Gordon, G. J. 1986. *Public Administration in America* (New York St. Martin's Press).

Gould, S. J. 1983. The titular bishop of Titiopolis. In *Hen's Teeth and Horse's Toes* (New York: W. W. Norton), 69–78.

Gouldner, A. W. 1954. *Patterns of Industrial Bureaucracy* (New York: Free Press).

Greer, A. L. 1981. Medical technology: Assessment, adoption, and utilization. *Journal of Medical Systems, 5*, 129–145.

Gupta, A., Raj, S. P., & Wilemon, D. 1986. A model for studying the

R&D–marketing interface in the product innovation process. *Journal of Marketing*, 50, 7–17.

Habermas, J. 1970. *Toward a Rational Society: Student Protest, Science, and Politics* (J. J. Shapiro, Trans.) (Boston: Beacon Press).

Habermas, J. 1981. *The Theory of Communicative Action: Vol 1. Reason and the Rationalization of Society* (Thomas McCarthy, Trans.) (Boston: Beacon Press).

Harvey, A. 1970. Factors making for implementation success and failure, *Management Science*, 16, B312–B321.

Hori, H. 1975. Revealed preferences for public goods. *American Economic Review*, 65(5), 978–991.

Hunt, S., & Nevin, J. R. 1974. Power in a channel of distribution: Sources and consequences. *Journal of Marketing Research*, 11, 186–193.

Hutchinson, S. E., & Sawyer, S. C. 1992. *Computers: The User Perspective* (New York: Irwin).

Huxhold, W. E. 1991. *An Introduction to Urban Geographic Information Systems* (New York: Oxford University Press).

Igbaria, M., & Nachman, S. A. 1990. Correlates of user satisfaction with end user computing: An exploratory study. *Information and Management*, 19, 73–82.

Imhof, E. 1963. Tasks and methods of theoretical cartography. *International Yearbook of Cartography*, 3, 13–23.

Ives, B., Olson, M. H., & Baroudi, J. J. 1983. The measurement of user information satisfaction. *Communications of the ACM*, 26, 785–793.

John, G. 1984. An empirical investigation of some antecedents of opportunism in a marketing channel. *Journal of Marketing Research*, 21, 278–289.

John, G., & Martin, J. 1984. Effects of organizational structure of marketing planning on credibility and utilization of plan output. *Journal of Marketing Research*, 21, 170–183.

Johnson, D. W. 1975. Cooperativeness and social perspective taking. *Journal of Personality and Social Psychology*, 31, 241–244.

Johnson, D. W., & Lewicki, R. J. 1969. The initiation of superordinate goals. *Journal of Applied Behavioral Science*, 5, 9–24.

Johnston, R. J., Gregory, D., & Smith, D. M. 1986. *The Dictionary of Human Geography* (2nd ed.) (Oxford, UK: Basil Blackwell).

Kaplan, B., & Duchon, D. 1988. Combining qualitative and quantitative methods in information systems research: A case study. *MIS Quarterly*, 12(4), 571–586.

Kasperson, R. E., & Pijawka, K. D. 1985. Societal response to hazards and major hazard events: Comparing natural and technological hazards. *Public Administration Review*, 45, 7–18.

Kearney, R. C., & Sinha, C. 1988. Professionalism and bureaucratic responsiveness: Conflict or compatibility. *Public Administration Review*, 48, 571–579.

Keller, L. F. 1984. The political economy of public management. *Administration and Society*, 15, 455–474.

Kellerman, A. 1983. Automated geography: What are the real challenges? *Professional Geographer*, 35, 342–343.

Kelly, H. H., & Stahelski, A. J. 1970. Social interaction basis of cooperators' and competitors' beliefs about others. *Journal of Personality and Social Psychology*, 16, 66–91.

Kirby, M. P. 1986. Politics, policy and computers in local government: Is there a role for the political scientist? *Urban and Regional Information Systems Association Conference Proceedings*, 4, 1–19.

Klosterman R. E. 1987. Guidelines for future computer-aided planning models. *Urban and Regional Information Systems Association Conference Proceedings*, 4, 1–14.

Knox, P. L. 1988. Disappearing targets? Poverty areas in central cities. *Journal of the American Planning Association*, 54, 501–508.

Kono, T. 1984. *Strategy and Structure of Japanese Enterprises* (Armonk, NY: M. E. Sharpe).

Kraemer, K. L., & Dutton, W. H. 1991. Survey research in the study of management information systems. In K. L. Kraemer (Ed.), *The Information Systems Research Challenge: Survey Research Methods* (Vol. 3) (Boston, MA: Harvard Business School), 3–58.

Kraemer, K. L., & King, J. L. 1976. *Computers, Power, and Urban Management: What Every Local Executive Should Know* (Beverly Hills: Sage).

Kraemer, K. L., King, J. L., Dunkle, D. E., & Lane, J. P. 1989. *Managing Information Systems: Change and Control in Organizational Computing* (San Francisco: Jossey-Bass).

Laughlin, P. R. 1978. Ability and group problem solving. *Journal of Research and Development in Education*, 12, 114–120.

Lawrence, P. R., & Lorsch, J. W. 1967. *Organization and Environment* (Homewood, IL: R. D. Irwin).

Lee, A. S. 1989. A scientific methodology for MIS case studies. *MIS Quarterly*, 13(1), 33–50.

Leonard-Barton, D. 1987. Implementing structured software methodologies: A case of innovation in process technology. *Interfaces*, 7(3), 6–17.

Liley, R. W. 1987. *Integration—The big payoff for municipal geo-based systems*. Paper presented at the annual Geographic Information Systems Conference, Falls Church, VA.

Lindahl, E. 1919/1958. Just taxation: A positive solution. In R. Musgrave & J. Peacock (Eds.), *Classics in the Theory of Public Finance* (London: Macmillan), 168–176.

Lipietz, A. 1986. New tendencies in the international division of labor:

Regimes of accumulation and modes of regulation. In A. J. Scott & M. Storper (Eds.), *Production, Work, Territory: The Geographical Anatomy of Industrial Capitalism* (Boston: Allen & Unwin), 16–40.

Lord, J. Dennis (Ed.). 1993. Open forum: Automated cartography in 1993 [Special issue]. *Professional Geographer, 45*(4), 431–460.

Lucas, H. C. Jr. 1975. Behavioral factors in system implementation. In R. L. Schultz & D. P. Slevin (Eds.), *Implementing Operations Research and Management Science* (New York: Elsevier), 203–216.

Lucas, H. C. Jr., & Nielson, N. R. 1980. The impact of the mode of information on learning and performance. *Management Science, 26,* 982–993.

Maguire, D. J., Goodchild, M. F., & Rhind, D. W. (Eds.). 1991. *Geographical Information Systems: Principles and Applications* (New York: Wiley).

Maital, S. 1973. Public goods and income distribution: some futher details. *Econometrica, 41,* 561–568.

Majone, G., & Wildavsky, A. B. 1978. Implementation as evolution. *Policy Studies Review Annual, 2.*

March, J. G. 1986. Theories of choice and making decisions. In F. S. Lane (Ed.), *Current Issues in Public Administration* (3rd ed.) (New York: St. Martin's Press), 229–247.

March, J. G., & Simon, H. A. 1958. *Organizations* (New York: Wiley).

Mayer, H. M. 1964, December. Politics and land use: The Indiana shoreline of Lake Michigan. *Annals of the Association of American Geographers, 54*(4), 508–523.

McCann, J., & Galbraith, J. R. 1981. Interdepartmental relations. In P. C. Nystrom & W. H. Starbuck (Eds.), *Handbook of Organizational Design* (Vol. 2) (New York: Oxford University Press), 60–84.

McGuire, M. 1974. Group segregation and optimal jurisdictions. *Journal of Political Economy, 80,* 112–132.

McLean, I. 1989. *Democracy and New Technology* (Cambridge, MA: Polity Press).

McNally, A. III. 1987. "You can't get there from here" with today's approach to geography. *Professional Geographer, 39,* 389–392.

Mead, D. A. 1981. Statewide natural-resource information systems: A status report. *Journal of Forestry, 79,* 369–372.

Meerman, J. 1978. Do empirical studies of benefit incidence make sense? *Public Finance, 3,* 295–313.

Merton, R. K. 1952. Bureaucratic structure and personality. In R. K. Merton, A. P. Gray, B. Hockey, & H. C. Selvin (Eds.), *Reader in Bureaucracy* (New York: Free Press), 361–372.

Metzendorf, D. 1988. An urgent need: Equal access of computers among the poor. In B. Glastonbury, W. LaMendola, & S. Toole (Eds.), *Information Technology and the Human Services* (New York: Wiley), 347–349.

Mayer, H. M. 1964. Politics and land use: The IN shoreline of Lake Michigan. *Annals of AAG, 54*, 508–523.

Miller, T. C. 1989. The operation of democratic institutions. *Public Administration Review, 49*, 511–521.

Milward, H. B. 1982. Interorganizational policy systems and research on public organizations. *Administration and Society, 13*, 457–478.

Mintzberg, H. 1979. *The Structure of Organizations* (Englewood Cliffs, NJ: Prentice-Hall).

Mitroff, I. I. 1975. On mutual understanding and the implementation problem: A philosophical case study of the psychology of the Apollo moon scientists. In R. L. Schultz & D. P. Slevin (Eds.), *Implementing Operations Research and Management Science* (New York: Elsevier), 237–252.

Moch, M., & Seashore, S. E. 1981. How norms affect behaviors in and of corporations. In P. C. Nystrom & W. H. Starbuck (Eds.), *Handbook of Organizational Design* (Vol. 1) (New York: Oxford University Press), 534–565.

Moenart, R. K., & Souder, W. E. 1990. An analysis of the use of extrafunctional information by R&D and marketing personnel: Review and model. *Journal of Product Innovation Management, 7*, 91–107.

Money, A., Tromp, D., & Wegner, T. 1988. The quantification of decision support benefits within the context of value analysis. *MIS Quarterly, 12*(2), 223–236.

Monmonier, M. 1991. *How to lie with maps* (Chicago: University of Chicago Press).

Mooney, J. D. 1947. *The Principles of Organization* (New York: Harper & Brothers).

Muehrcke, P. C. 1981. Whatever happened to geographic cartography? *Professional Geographer, 33*, 397–405.

Musgrave, R., & Peacock, A. T. (Eds.) 1958. *Classics in the Theory of Public Finance* (New York: Macmillan).

Newman, W. H. 1988. *Role of departments in multi-level strategic management* (Working Paper No. SC 65) (New York: Columbia University, Strategy Research Center).

Niehoff, A. H. 1966. *A Casebook of Social Change* (Chicago: Aldine).

North, R. C., Koch, H. E. Jr., & Zinnes, D. 1960. The integrative functions of conflict. *Journal of Conflict Resolution, 4*(3), 355–374.

O'Dowd, D. J. 1988. The relevance of information technology for policing: Dilemmas of control and freedom. In R. Plant, F. Gregory, & A. Brier (Eds.), *Information Technology: The Public Issues* (London: Manchester University Press), 148–163.

Oakland, W. H. 1972. Congestion, public goods and welfare. *Journal of Public Economics, 1*(3), 339–357.

Obermeyer, N. J. 1990a. *Bureaucrats, Clients and Geography: The Bailly*

Nuclear Power Plant Battle in Northern Indiana (Research Paper No. 216) (Chicago: University of Chicago, Department of Geography).

Obermeyer, N. J. 1990b. Regional equity in turbulent times: The experience of the Regional Transportation Authority of northeastern Illinois. *Applied Geography, 10,* 147–161.

Obermeyer, N. J. 1991, February. Sharing geographic information across organizational boundaries: An organizational–managerial perspective. *Proceedings of GIS '91,* 132–141.

Oldham, G. R., & Brass, D. J. 1979. Employee reactions to an open-plan office: A naturally occurring quasi-experiment. *Administrative Science Quarterly, 24,* 267–284.

Olson, M. 1971. *The Logic of Collective Action: Public Goods and the Theory of Groups* (Cambridge, MA: Harvard University Press).

Olson, M., & Zeckhauser, R. 1969. An economic theory of alliances. *Review of Economics and Statistics, 48,* 266–279.

Onsrud, H. J. 1989. Legal and liability issues in publicly accessible land information systems. *Proceedings of GIS/LIS '89, 1,* 295–300.

Onsrud, H. J. 1992. In support of open access for publicly held geographic information. *GIS Law, 1*(1), 3–6.

Onsrud, H. J., Calkins, H. W., & Obermeyer, N. J. (Eds.). 1989. *Use and value of geographic information: Initiative four specialist meeting* (Tech. Paper No. 89-7) (Santa Barbara: University of California, National Center for Geographic Information and Analysis).

Onsrud, H. J., & Hintz, R. J. 1991. Evidentiary admissibility and reliability of automated field recorder data. *Surveying and Land Information Systems, 51,* 23–28.

Onsrud, H. J., & Pinto, J. K. 1991. Diffusion of geographic information innovations. *International Journal of Geographic Information Systems, 5,* 447–467.

Onsrud, H. J., & Pinto, J. K. 1992. *Correlating adoption factors and processes with GIS user satisfaction in U.S. local governments.* Paper presented at NATO Advanced Research Workshop on Diffusion of Geographic Information Technologies, Sounion, Greece.

Onsrud, H. J., Pinto, J. K., & Azad, B. 1992. Case study research methods for geographic information systems. *URISA Journal, 4,* 32–44.

Orwell, George. 1949. *1984* (New York: Harcourt, Brace).

Papageorgiou, G. J. 1978, December. Spatial externalities: Parts I and II. *Annals of the Association of American Geographers, 68*(4), 465–492.

Parker, H. D. 1987. *What Is a Geographic Information System?* (Falls Church, VA: American Society for Photogrammetry and Remote Sensing, in association with American Congress on Surveying and Mapping).

Pavett, C. M., & Lau, A. W. 1983. Managerial work: The influence of hierarchical level and functional specialty. *Academy of Management Journal, 26,* 170–177.

Perry, J. L., & Kraemer, K. L. 1979. *Technological Innovation in American Local Governments_the Case of Computing* (New York: Pergamon Press).

Peters, T. 1990. Get innovative or get dead. *California Management Review, 33*(1), 9–26.

Pfeffer, J. 1982. *Organizations and Organization Theory* (Boston: Pitman).

Pfeffer, J. & Salancik, G. R. 1978. *The External Control of Organizations* (New York: Harper & Row).

Pfiffner, J. P. 1987. Political appointees and career executives: The democracy–bureacracy nexus in the third century. *Public Administration Review, 47*, 57–65.

Pickles, J. 1993. Discourse on method and the history of discipline: Reflections on Dobson's 1983 automated geography. *Professional Geographer, 45*(4), 451–455.

Pinto, M. B., Pinto, J. K., & Prescott, J. E. 1993. Antecedents and consequences of project team cross-functional cooperation, *Management Science, 39*, 1281–1297.

Piore, M. J. 1979. Qualitative research techniques in economics. *Administrative Science Quarterly, 24*, 560–569.

Pondy, L. R. 1967. Organizational conflict: Concepts and models. *Administrative Science Quarterly, 12*(2), 296–320.

Porter, M. E. 1979. How competitive forces shape strategy. *Harvard Business Review, 57*(2), 137–145.

Portner, J., & Niemann, B. J. Jr. 1983. Belief differences among land records officials and users: Implications for land records modernization. *Proceedings of the Urban and Regional Information Systems Association*, Washington, DC.

Pressman, J. L., & Wildavsky, A. B. 1973. *Implementation* (Berkeley: University of California Press).

Prisley, S. P., & Mead, R. A. 1987. Cost-benefit analysis for geographic information systems. *Proceedings of GIS '87*, Falls Church, VA.

Pugh, D. L. 1989. Professionalism in public administration: Problems, perspectives, and the role of ASPA. *Public Administration Review, 49*, 1–8.

Quinn, J. B. 1980. *Strategies for Change* (Homewood, IL: Irwin).

Raghavan, S. A., & Chandf, D. R. 1989. Diffusing software-engineering methods. *IEEE Software, 15*(7), 81–90.

Raia, A. P. 1974. *Managing Objectives* (Glenview, IL: Scott Foresman).

Rasmussen, P. W. 1986. What should we do with the AICP exam? *Journal of the American Planning Association, 52*, 7–8.

Raymond, L. 1987. Validating and applying user satisfaction as a measure of MIS success in small organizations. *Information and Management, 12*, 173–179.

Reukert, R. W., & Churchill, G. A. Jr. 1984. Reliability and validity of alternative measures of channel member satisfaction. *Journal of Marketing Research, 21*, 226–233.

Reukert, R. W., & Walker, O. C. 1987. Marketing's interaction with other functional units: Conceptual framework and empirical evidence. *Journal of Marketing, 51,* 1–19.

Rice, R. E., & Rogers, E. M. 1980. Re-invention in the innovation process. *Knowledge: Creation, Implementation, Utilization, 1,* 499–514.

Rogers, E. M. 1962. *Implementation of Innovations* (New York: Free Press).

Rogers, E. M. 1983. *Implementation of Innovations* (3rd ed.) (New York: Free Press).

Rogers, E. M., & Shoemaker, F. F. 1971. *Communication of Innovations: A Cross-Cultural Approach* (New York: Free Press).

Roitman, H. 1988. Public records laws: A proposed model for changes. *Proceedings of the 26th Annual Urban and Regional Information Systems Association, 4,* 338–347.

Saarinen, A. O. 1987. Improving information systems development success under different organization conditions. *Urban and Regional Information Systems Association Conference Proceedings, 11,* 1–12.

Samuelson, P. A. 1954. The pure theory of public expenditures. *Review of Economics and Statistics, 36,* 387–389.

Sayer, A. 1986. Industrial location on a world scale: The case of the semiconductor industry. In A. J. Scott & M. Storper (Eds.), *Production, Work, Territory: The Geographical Anatomy of Industrial Capitalism* (Boston: Allen & Unwin), 107–124.

Schermerhorn, J. R. 1975. Determinants of interorganizational cooperation. *Academy of Management Journal, 18,* 846–856.

Schermerhorn, J. R. Jr. 1989. *Managing for Productivity* (New York: Wiley).

Schmidt, W. H., & Tannenbaum, R. 1960. The management of scientific manpower. *Management Science, 14,* B473–B489.

Schultz, R. L., & Henry, M. D. 1981. Implementing decision models. In R. L. Schultz & A. A. Zoltners (Eds.), *Marketing Decision Models* (New York: North-Holland), 123–134.

Schultz, R. L., Ginzberg, M. J., & Lucas, H. C. Jr. 1983. A structural model of implementation (Working Paper), University of Texas, Austin.

Schultz, R. L., & Slevin, D. P. 1975. Implementation and management innovation. In R. L. Schultz & D. P. Slevin (Eds.), *Implementing Operations Research/Management Science* (New York: Elsevier), 3–20.

Schultz, R. L., & Slevin, D. P. 1979. Introduction: The implementation problem. In R. Doktor, R. L. Schultz, & D. P. Slevin (Eds.), *The Implementation of Management Science* (New York: North-Holland), 1–15.

Schultz, R. L., Slevin, D. P., & Pinto, J. K. 1987. Strategy and tactics in a process model of project implementation. *Interfaces, 17*(3), 34–46.

Scott, B. W. 1963. *Some aspects of long-range planing in American corporations with special attention to strategic planning.* Unpublished doctoral dissertation, Harvard University, Cambridge, MA.

Shapiro, B. P. 1977. Can marketing and manufacturing coexist? *Harvard Business Review*, 55, 104–114.

Shapiro, E. 1993, December 16. Cigarette maker and Time aim ads at smokers. *Wall Street Journal*, B1, B7.

Sheppard, E. 1993. Automated geography: What kind of geography for what kind of society. *Professional Geographer*, 45(4), 457–460.

Sherif, M. 1962. Intergroup relations and leadership: Introductory statement. In M. Sherif (Ed.), *Intergroup Relations and Leadership* (New York: Wiley), 3–21.

Sherif, M., & Sherif, C. W. 1969. *Social Psychology* (New York: Harper & Row).

Simon, H. A. 1945/1976. *Administrative Behavior: A Study of Decision-making Processes in Administrative Organization* (New York: Free Press).

Simon, H. A. 1952. Decision-making and administrative organization. In R. K. Merton, A. P. Gray, B. Hockey, & H. C. Selvin (Eds.), *Reader in Bureaucracy* (New York: Free Press), 185–193.

Simon, H. A. 1964. On the concept of organizational goals. *Administrative Science Quarterly*, 9, 1–22.

Smith, C. 1981. Urban structure and the development of narual support systems for service-dependent populations. *Professional Geographer*, 33, 547–565.

Smith, D. A., & Tomlinson, R. F. 1992. Assessing the costs and benefits of geographical information systems: Methodological and implementation issues. *International Journal of Geographical Information Systems*, 6(3), 247–256.

Smith, N. 1992. History and philosophy of geography: Real wars, theory wars. *Progress in Human Geography*, 16(2), 257–271.

Souder, W. E. 1981. Disharmony between R&D and marketing. *Industrial Marketing Management*, 10, 67–73.

Souder, W. E. 1988. Managing relations between R&D and marketing in new product development projects. *Journal of Product Innovation Management*, 5, 6–19.

Srinivasan, A., & Davis, J. G. 1987. A reassessment of implementation process models. *Interfaces*, 17(3), 64–71.

Steinbruner, J. D. 1974. *The Cybernetic Theory of Decision* (Princeton, NJ: Princeton University Press).

Stern, L. W., & Heskett, J. L. 1968. Conflict management in interorganizational relations: A conceptual framework. In L. W. Stern (Eds.), *Distribution Channels: Behavioral Dimensions* (Boston, MA: Houghton Mifflin), 288–305.

Stern, L. W., Heskett, J. L., Sternthal, B., & Craig, C. S. 1973. Managing Conflict in distribution channels: A laboratory study. *Journal of Marketing Research*, 10, 169–179.

Stough, R. R., & Whittington, D. 1985. Multijurisdictional waterfront land use modeling. *Coastal Zone Management Journal*, *13*, 151–175.

Street, J. 1988. Taking control? Some aspects of the relationship between information technology, government policy and democracy. In R. Plant, F. Gregory, & A. Brier (Eds.), *Information Technology: The Public Issues* (London: Manchester University Press), 148–163.

Taylor, F. W. 1911. *Principles of Management* (New York: Harper & Brothers).

Thomas, K. 1976. Conflict and conflict management. In M. D. Dunnette (Ed.), *Handbook of Industrial and Organizational Psychology* (Chicago, IL: Rand McNally), 889–935.

Thompson, J. D. 1967. *Organizations in Action* (New York: McGraw-Hill).

Thompson, J. D. 1969. *Bureaucracy and Innovation* (University, AL: University of Alabama Press).

Thompson, A. A. Jr., & Strickland, A. J. III, 1987. *Strategic Management: Concepts and Cases* (4th ed.) (Plano, TX: Business Publications).

Thompson, V. A. 1976. *Bureaucracy and the Modern World* (Morristown, NJ: General Learning Press).

Tiebout, C. M. 1956. A pure theory of local expenditures. *Journal of Political Economy*, *64*, 416–424.

Tjosvold, D. 1984. Cooperation theory and organizations. *Human Relations*, *37*, 743–767.

Tobler, W. R. 1970. A computer movie simulating urban growth in the Detroit region. *Economic Geography*, *46*, 234–240.

Tobler, W. R. 1976. Analytical cartography. *American Cartographer*, *3*(1), 21–31.

Tomlinson, R. F. 1985. Geographic information systems: The new frontier. *Operational Geographer*, *5*, 31–36.

Trist, E. 1977. Collaboration theory and organizations. *Journal of Applied Behavioral Science*, *13*, 268–278.

Trubow, G. B. 1991. An overview of information technology and privacy in the United States. *World Computer Law Congress*.

Urry, J. 1986. Capitalist production, scientific management and the service class. In A. J. Scott & M. Storper (Eds.), *Production, Work, Territory: The Geographical Anatomy of Industrial Capitalism* (Boston: Allen & Unwin), 41–66.

Van de Ven, A. H., Delberg, A. L., & Koenig, R. Jr. 1976. Determinants of coordination modes within organizations. *American Sociological Review*, *41*, 322–338.

Voltaire. 1981. *Candide* (L. Bair, Trans.) (Toronto, Ontario: Bantam Books).

Wall, J. A. Jr. 1985. *Negotiation Theory and Practice* (New York: Scott, Foresman).

Walton, R. E., & Dutton, J. M. 1969. The management of interdepartmental conflict: Model and review. *Administrative Science Quarterly*, *14*, 73–84.

Wang, M., & Stough, R. R. 1986. Cognitive analysis of land-use decision-making. *Modeling and Simulation: Part 1. Geography—Regional Sciences, Economics* (Proceedings of the 17th annual Regional Science Association Conference in Pittsburgh), *17*, 107–112.

Weber, M. 1946. Bureaucracy. In H. H. Gerth & C. W. Mills, (Eds. and Trans.), *From Max Weber: Essays in Sociology* (New York: Oxford University Press), 196–244.

Weber, M. 1968. Sociological categories of economic action. In G. Roth & C. Wittich (Eds. and Trans.), *Economy and Society* (Vols. 1 & 2) (Berkeley: University of California Press), 63–210.

Webster's II New Riverside University Dictionary. 1984. (Boston: Riverside).

Wellar, B. 1988a. Institutional maxims and conditions for needs-sensitive information systems and services in local governments. *Proceedings of the 26th Annual Conference of the Urban and Regional Information Systems Association, 4,* 371–378.

Wellar, B. 1988b. A framework for research on research in the information technology—local government field. *Proceedings of the 26th Annual Conference of the Urban and Regional Information Systems Association, 4,* 379–386.

Wentworth, M. J. 1989. Implementation of a GIS project in a local government environment: The long and winding road. *Proceedings of the 27th Annual Urban and Regional Information Systems Association, 2,* 198–209.

Whiteman, J. 1983. Deconstructing the Tiebout hypothesis. *Environment and Planning—D: Society and Space, 1*(3), 339–354.

Wiggins, L. L., & French, S. P. 1991. In press. *Geographic Information Systems: Assessing Your Needs and Choosing a System* (Planning Advisory Service Report) (American Planning Association, Chicago).

Wilcox, D. L. 1990. Concerning "The economic evaluation of implementing a GIS." *International Journal of Geographical Information Systems, 4*(2), 203–210.

Wilke, J. R. 1993, December 9. Shop talk: Computer links erode hierarchical nature of workplace culture. *Wall Street Journal,* A1, A7.

Williams, A. 1966. The optimal provision of a public good in a system of local government. *Journal of Political Economy, 74,* 18–33.

Williams, F., Rice, R. E., & Rogers, E. M. 1988. *Research Methods and the New Media* (New York: Free Press).

Williams, R. E. 1987. *Selling a geographical information system to government policy makers.* Paper presented at the annual conference of the Urban and Regional Information Systems Association, Washington, DC.

Williamson, O. E. 1975. *Markets and hierarchies: Analysis and antitrust implications.* New York: Free Press.

Wunderlich, G. 1986. *Land use information systems for rural jurisdictions.* Paper presented at the annual conference of the Urban and Regional Information Systems Association, Washington, DC.

Yellin, J. 1981. High technology and the courts: Nuclear power and the need for institutional reform. *Harvard Law Review, 94*(3), 489–560.

Zaltman, G., & Duncan, R. 1977. *Strategies for Planned Change* (New York: Wiley).

Zaltman, G., & Moorman, C. 1989. The management and use of advertising research. *Journal of Advertising Research, 28*, 11–18.

Zand, D. E. 1972. Trust and managerial problem solving. *Administrative Science Quarterly, 17*, 229–239.

Zelinsky, W. 1980. North America's vernacular regions. *Annals of the Association of American Geographers, 70*, 1–16.

Zwart, P. 1991, August. *Some indicators to measure the impact of land information systems in decision making.* Paper presented at the annual conference of the Urban and Regional Information Systems Association, San Francisco.

❖

Index